Ham and 2-Way Radio

Common Frequencies Reference Guide

A Reference Guide for Radio Preppers, Hams, and Enthusiasts with popular Shortwave and common 2-way radio frequency listings

By Rodney Biddle

KX4HD, Amateur Extra

http://www.KX4HD.com

Ham and 2-Way Radio – Common Frequencies Reference Guide

ISBN-13: 978-0-9996491-0-7

The web addresses cited in this book were current as of November 2017, unless otherwise noted.

Disclaimer:
The information given in this book is solely for educational and entertainment purposes and in no way should be construed to be legal, medical or professional advice. The author and publisher assume no liability for any actions taken by any individual or group that is considered illegal, based on information in this book.

Table of Contents

Forward

I had just wrapped up going through a home show with a friend up in Atlanta a few years ago. Professionally I am a web developer with a focus towards doing online publications. My friend had asked me for some guidance on a project he was working on and we visited the home show for some ideas about his topic. Afterwards he had somewhere to be and I had a few hours to kill, and there was a Prepper show in the hall next door. I was looking for some new ideas and topics for my writing and thought this could be an interesting show to visit.

One of the exhibitors was discussing radio communications and emergency planning during a catastrophic event. The discussion turned to communicating after the loss of power and the types of events that could create such an event. The worst-case picture that was being painted was the loss of the US power grid – a topic that has been more in the news with EMP discussions, hacker threats, and the real possibility of a solar flare. (Google "Carrington Event 1859" or "July 2012 Solar Flare"). The more common causes though can be seen just from recent 2017's Hurricane season. Hurricane Harvey, Irma and Maria have knocked out cellular systems, electrical grids, and traditional telephone lines in the United States and Puerto Rico. Ham radio volunteers working in Puerto came together to help thread back together a devastated communications network bringing help to remote areas cut off.

In college, I graduated with a degree in Electronics and had even obtained my first radio license at that time – I have held a Commercial Radio Telephone license for over 30 years. But did not stay with the hobby into my later career years. I decided to roll up my sleeves and dig back into the topic, first getting my Technician license and shortly after following up with my General and Amateur Extra licenses obtaining my call sign with my Extra-level license of KX4HD.

On my journey to obtain my license I learned about the different radio technologies and services gaining a clearer understanding through use and study of Ham radio. I also came to find a lack of good beginner materials that covered the various radio services for the beginner in a consolidated format. There are plenty of sources for information on the services, but harder to find in a single location. Also difficult to find was a consolidated reference to the frequencies available to use and which services can cross communicate if necessary.

Thus, this book. My goal with this book is to provide some guidance and focus when it comes to radio technology. What radio systems to use – what their strengths are, limitations, and legal use. Getting licensed and the benefits of licensing. This is not a buyer's guide and I am not recommending specific radios, but we will work through the technologies and provide our readers a list of important radio features and frequencies to have on hand if you need. There are so many frequencies and they vary by area, state or region. We have published nationally established frequencies you can reference however including frequencies used on Ham, GMRS, FRS, MURS, Citizens Bands, and national government and military channels.

Use this book as a guide to help you decide the technology you are looking for and the license levels needed (If any). Licensing is not a big thing – You can get a GMRS License inexpensively with no test that covers all of your family. Or an inexpensive Ham license for yourself (Testing necessary) that will expand your capabilities on radio. Or select a license-free method such as MURS, FRS or Citizens Band (CB) but understand the limitations.

Decide on a radio which also can be cheap – under $100 (or less). Learn your radio, and keep this book handy for reference to frequencies.

If a worse case event ever does happen, you will need to know a lot to survive. Radio will be a useful tool for survival for yourself or your family. I hope this book can become a useful and relied upon resource for you to add to your library.

1- Introduction

Ham radio draws the attention of those of us who either have a passion for the interest in talking to new people around the country or around the world, curious to see who we will reach and where, or it also tends to draw the interest in those of us wanting to use radio as a utility for communications. Whether it be to communicate where our cell phones simply do not work, or while backpacking in remote areas, or for use during emergencies as backup communication, radio is a critical tool that we use and has become, to some extent, a lesser used and known skill with the popularity of cellular phones.

For the newcomer to radio it can be an overwhelming question as to what types of radios and frequencies are available. It is even more overwhelming to the newcomer who has an interest in radio and may want to learn what capabilities are available to them at various levels. License free radio is readily available in the forms of Citizens Band, FRS and MURS radio. Each system has its own benefits and limitations. Moving up the communications ladder we move into higher powered radio capabilities such as GMRS and Amateur radio, both of which require a level of licensing to use. GMRS, while often limited by the FRS/GMRS one-radio-for-all packages, does can support higher power of up to 50 watts and repeaters.

If your interest is in moving up to Amateur radio you can pursue one of three levels of licenses, each of which grants progressive permissions in the various amateur bands. Getting into amateur radio at the base levels opens the legal use of the 2 meter, 1.25 meter and 70 cm bands, limited access to other bands, access to repeater networks, and legal use of radio frequencies using base, handheld and mobile radios. Higher license levels open more bands including higher meter bands (6, 10, 20, 40 and other bands) and more access to frequencies in those bands that open opportunities to be able to reach out much greater distances with more power extending hundred and thousands of miles across the country and across the world.

Early in our book we are going to touch on various radio systems available. From shortwave radio for listening to world broadcasts, then we will look at license free radio systems and to the more advanced and powerful different license based radio systems. We will get into the meat of what this book is about in the later portion of the book – indexes with frequencies of public and government national service frequencies. We also will include links to resources available online and in print useful to have available for emergency planning, and references to state and regional resources that can provide more established frequency information including repeater networks that are local to each state or region.

We must emphasize that the use of many of the frequencies listed in this book are not for use by the public. The public safety, government, law enforcement, air band and other channels that we have in here for reference are strictly reserved for those individuals and agencies with a direct need for those frequencies. They can be listened to and monitored which is why we have listed them here for reference. If an emergency should occur, then having these available for monitoring could help you to stay informed on news in your area if regular communications are not available.

The frequencies that we have as reference in our book cannot be accessed by every type of radio. As with tools, each radio system has its own use and capabilities. For instance, CB Radios pick up a limited range of CB Channels, as do MURS Radios with MURS Frequencies, and FRS with FRS Frequencies. With FRS and GMRS there is a caveat that since they share many or all the same channels, many of these radios can communicate with each other although range will vary because of power and antenna differences. Many frequencies will require specialized radios or scanners which are listen-only radios specifically designed for tuning into a wider range of radio frequencies, though not all frequencies.

2 - Public Two-Way Radios & Frequencies

Of the types of radio systems popularly available to the public for everyday use, we are looking at the most common for public general use in this book. These are easily available with two requiring licensing and three license free solutions.

FRS (Family Radio Service) Radio

FRS is a low-power low-capability radio system designed for short distance license free communication. In May 2017, the FCC made a move to extend the power of FRS radio on some frequencies to 2 watts of power, but still mandates restrictions on antenna design which makes these radios good for short-distance use but not viable beyond and eighth to a quarter mile (You might get a half mile with no obstacles). New radios with the new power limits are yet to be available as of the time of this book in summer 2017 but should be hitting the market soon. Restrictions on the antenna requires the antenna to be permanently fixed to the radio which is the biggest limitation of these radios.

FRS/GMRS Combination radios will stop being certified by the FCC in 2017 or 2018 and new models which no longer are mixed use FRS/GMRS radios will become available at some point.

FRS in a Snapshot:

- No License Required
- Low Power (Max 2 Watts with newer radios made after 2017/2018)
- Short Range – usually less than 1 Mile
- 22 Fixed channels with ½ watt or 2-watt power depending on channel, max of 2 watts of power.
- Fixed antenna – No external antenna options available

GMRS (General Mobile Radio Service)

GMRS is a step up from FRS with power levels up to 5 watts for hand-held radios and the ability to use external antennas as well as repeaters (External antennas are available only on dedicated GMRS Radios – Not combo radios). GMRS radio does require a license for legal use, but the license does not require a test.

The maximum range for GMRS with repeaters and an external antenna can easily extend to 10 miles or more depending on numerous factors, though the common FRS/GMRS Radios found in retailers will rarely get more than a mile or two depending on the area they are being used. These radios operate in the UHF radio bands in the 460 MHz range where the frequencies are "Line-of-site". Frequencies of this range are often blocked by obstacles and terrain which becomes their limiting factor and the reason an external antenna can be so beneficial to these. It is possible for two GMRS radio operators to communicate 20 miles or more with the right conditions.

Most common GMRS radios are packaged as "FRS/GMRS" combo radios. This combination packaging requires the antennas be permanently fixed to the radio though many of these radios will automatically switch power depending on if you are using an FRS or GMRS frequency.

A GMRS radio license is required for use which costs about $70 and lasts for 10 years, good for all members of the purchaser's family.

GMRS in a Snapshot:

- License IS Required - $70 for a 10-year license (Covers all family members)
- More Power (Max 5 Watts handheld, 50 watts on a mobile radio
- Better Range – Several miles depending on power, antenna and obstacles
- Radio-to-Radio range usually less than 5 miles (5 Watts, no repeater)
- 22 channels with ½ watt, 2 watts, 5 watts or 50 watts of power depending on channel.
- External antenna options available extending range capabilities.

Citizens Band "CB" Radio

Citizens Band Radio exploded into public back in the 70's before cellular telephones. It quickly became popular among truckers and drivers on the highways to communicate. It operates in the 27 MHz frequency range requiring much longer antennas than on other radio types making it less convenient for handheld use and more common for vehicles. Power limits are 4 watts for normal use and 12 watts for single-side-band use giving it a rough range, depending on antenna and conditions of between 5 and about 15 miles.

Antennas are generally several feet in length and external to the radio, usually mounted on a vehicle or outside. A full quarter-wave antenna for a hand-held radio would extend to over 9 feet in length which is why almost all hand-held versions of these radios use a less efficient shorter antenna.

No license is required for CB Radio.

CB Radio in a Snapshot:

- No License Required
- Low Power (Max 4 Watts *Legal* standard channels or 12 watts on Single Side Band channels.)
- Medium Range – usually less than 10 Miles depending on Antenna
- 40 Fixed channels 80 additional "Sideband" channels
- External antenna options available. Range is highly dependent on Antenna. Antenna lengths are long – roughly 8.5 feet for a ¼ wave. Shorter antennas will usually shorten range.

What about High Power CB? There are CB'ers with a lot more than 12 watts of power?

The references we are covering are for the LEGAL power limits of CB Radio. There are a lot of folks in the hobby who push well beyond these legal limitations with 100 watts or more power. The FCC Legal limits however are 4 watts full modulation, 12 watts SSB modulation.

MURS Radio service

MURS is a 5-channel radio system that was carved out of the Business radio systems in 2000. The frequencies used are in the 151 and 154 MHz frequency ranges of VHF Radio and these radios work at 2 watts of power. The ability to utilize an externally connected antenna allows better communication signal but at 2 watts these still are more limited than a dedicated GMRS radio with an external antenna running at 5 watts. MURS radios are NOT allowed to use repeaters. Operation will be better however than an FRS radio even under the new FCC Limits of 2 watts for FRS due to the ability to use an externally attached antenna.

In areas where businesses still hold older licenses for MURS frequency use those businesses do take precedence over public use on MURS channels.

MURS in a Snapshot:

- No License Required
- Low Power (Max 2 Watts)
- Short Range – usually less than 2 Miles depending on antenna
- 5 Fixed channels with 2-watts of power
- External antenna options available

Amateur Ham Radio

This radio is made up of a range of frequencies throughout the spectrum available for use by licensed ham operators depending upon purpose and license level. Ham radio operators are licensed in one of three levels ranging from the entry-level Technician's license, the General license and highest level of Amateur Extra license. Each progressive level of licensing allows access to more frequencies and in different radio bands ranging from 135 kHz at the bottom up to as high as 275 GHz, but only narrow blocks of frequencies within the various bands and not for all license levels. As a Technician, the entry level license holder, the most popular bands are the VHF 2-Meter band and the UHF 70-cm band on which voice communication is allowed.

With the introduction of inexpensive Chinese made radios in recent years and a growth in the Prepper community to learn emergency radio communications, a growth in the 2-Meter and 70-cm communications and ham operators becoming licensed at the technician level has increased traffic in these bands. The VHF (2 Meter) and UHF (70-cm and 1.25-Meter) frequencies are generally used as line-of-site systems with typical radio ranges of a few miles, but support with external antennas and use of radio repeater networks allows the ham operators in these frequencies to communicate dozens of miles reaching distances of a hundred miles or more depending upon the repeater network layout. Repeaters installed and extreme heights allow low-power radios to communicate much greater distances, and when tied together then a user at one radio repeater can communicate easily to other users tied in with other repeaters in the network.

Ham operators using lower HF frequencies can take advantage of radio wave propagation to communicate directly with other hams hundreds or thousands of miles without the need for repeaters. These ham operators communicate by doing so on frequencies where their radio signals are bouncing through the atmosphere to other operators at great distances.

Getting licensed as a ham operator takes study and testing. The cost is around $15 or so per test but the operator does need to show understanding of the content at various levels.

Amateur Radio in a Snapshot:

- License Required – 3 Levels (Technician, General and Amateur Extra)
- Higher Power – Handhelds from 5 watts up to stationary allowing up to 1500 watts of power. Common power for home-base radios is 100 watts.
- Long Range – Hand-held point-to-point of a few miles to thousands of miles with full-power radio units.
- Most tuning is performed by frequency with established frequencies published for general call frequencies.
- Wide range of antenna options available which allow wide range of capabilities.
- Portable hand-held radios can be purchased for roughly $30 (70cm/2-meter) up to hundreds or thousands of dollars depending on capabilities.

Marine Radio and Airband (Aircraft) Radio

Both of these radio systems utilize fixed channels with restrictions on both systems. Due to the restrictions we will only be touching on these radio systems. Both are legally restricted in use though no license is specifically required as long as they are used as intended. Marine radio only for on-water boat use, Airband radio for aircraft communication.

Publicly available radio frequencies available for use with the appropriate license. Note that FRS, CB and MURS do not require licensing however do have power and antenna restrictions which must be followed for legal use. The following chart covers several radio systems readily available to the public for two-way communication that we are providing frequency information for.

	FRS	GMRS	MURS	CB	Amateur
Frequency Range	UHF 462/467 MHz	UHF 462/467 MHz	VHF 151 MHz	HF 29 MHz	Multiple Bands - Based on License (1)
Max Power	1/2 Watt	5 Watts Handheld, 50 Watts Mobile	2 Watts	4 Watts Normal (Max 12 Watts SSB)	1,500 Watts (6)
Antenna Type	Permanent Fixed	External (2)	External	External	External
Repeaters	No	Yes (3)	No	No	Yes
Est Range w/o Repeater	½ Mile to 1-2 miles depending on surroundings.	Est 1-3 Miles, 10+ Miles w/Ext Antenna	Est 2-3 Miles Up to 10 Miles w/Ext Antenna	Est 5-10 Miles	Hundreds to Thousands of Miles (4)
Est Range w/Repeaters	NA	~ 50 Miles	NA	NA	Dozens to 1000's (5)
Mode(s)	FM	FM	AM, FM	AM, SSB	AM, FM, SSB, CW, Digital, VOIP
License Needed	No	Yes - Family License	No	No	Yes 3-Levels
License Cost	None	~ $70	None	None	~ $15 per test

(1) Frequencies and bands are dependent upon the level of the Amateur license.

(2) GMRS Radios which are combination FRS/GMRS must use a fixed attached antenna. Otherwise, dedicated GMRS Radios can use a replicable external antenna allowing much greater range.

(3) Repeaters with limitations can be used for GMRS on specific GMRS Frequencies.

(4) Distance on Amateur radio depend upon antenna, power, and frequencies. Long distances are common on lower frequency bands, and limited on Higher frequency bands (HF, VHF, UHF) without repeaters.

(5) Repeaters are more common with VHF & UHF Frequencies, and necessary for many radio types. Handhelds can only communicate long distances over VHF/UHF with repeaters which can be hundreds of miles.

(6) Power levels dependent upon license and band being used.

These five radio systems are the most readily available for the general public for common land use. Marine Radio is also an available option for public use however is not legal for use on land. Know also that not all of the frequencies provided in the chart on the previous page can be received by all radios. Not all radios are similar and just as with tools – you must have the right one for the right job.

The most versatile of the radio types is by far the Ham Radio which can be tuned in, depending on the model of the radio, to most of these channels. A popular dual-band handheld HAM radio such as the Baofeng UV-5R, Yaesu hand radios, or many Kenwood dual band radios can pick up HAM channels, FRS, MURS, and GMRS Frequencies whereas most others are limited to their own bands.

Important note: Though you may receive a channel on a particular type of radio without a license, this does not mean that transmitting on those same channels or frequencies will be legal. As with tools however each type of radio has its own purpose. Transmitting without a license and/or causing interference can result in fines from the FCC in thousands of dollars. The FCC publishes a list of headlines including violations on their website at http://transition.fcc.gov/eb/

The goal of this book is to provide guidance and reference to the radio operator of common and popular national frequencies as well as a reference to more local resources where you can obtain local frequency and channel information that is more specific to your area. We are providing in later sections references to repeater directories as well as regional and state repeater councils where you can find local repeater networks. This book does not delve deeper into the radios themselves or legal use of specific radio systems. We are strictly focused on providing frequency and channel reference details as well as some basics on reception and other references that are useful for the radio enthusiast and the newcomer to radio.

3 - Shortwave Radio and Directories

Shortwave radio is the longest reaching radio system which you can use to tune into with specialized equipment specific for shortwave frequencies. With the explosion of the Internet, Shortwave radio has gone through a reduction in popularity in many parts of the world including the United States. During an emergency, however where power and the Internet may not be available, radio will still reign as the only tool that we can rely on. With Shortwave radio being able to reach thousands, or even tens of thousands of miles, it will be the most likely system for receiving news from the longest distances if we were to have a wide-reaching catastrophe reaching across large areas or regions, Shortwave Radio will be an essential method of receiving news.

Below are several popular frequencies that are on at various times of the day. Note through that not all stations broadcast 24 hours.

Country:	Frequencies:
ALASKA	6150, 6950, 7355, 9920
CANADA RCI	9610, 9755, 9770, 13650, 15365, 17740
CANADA CBC	6160, 9625
FRANCE	5920, 7315, 9720, 9765, 9805, 9865, 11615, 11725, 13680, 11995, 15160, 15275, 15605, 21620
GERMANY	5905, 6140, 6180, 7225, 7240, 7280, 7285, 9565, 9735, 9755, 11690, 12045, 15275
MEXICO	6185
UNITED KINGDOM	5875, 5975, 6005, 6040, 6130, 6195, 7130, 7160, 7320, 9410, 9480, 9660, 9740, 9750, 11675, 11750, 11765, 11920, 12095, 15105, 15285, 15360, 15400, 15575, 17640, 17830, 17885, 21470
UNITED NATIONS	9565, 17810
USA AFRTS	4319-USB, 5446.5-USB, 5765-USB, 6350-USB, 7811.5-USB, 10320-USB, 12133.5-USB, 12759-USB, 13362-USB
USA VOA	4930, 4960, 5960, 6080, 6105, 6110, 7125, 7175, 7205, 7405, 9645, 9760, 9785, 9885, 11655, 11885, 11890, 11975, 12015, 12150, 13600, 13640, 13710, 13735, 13755, 15150, 15185, 15205, 15290, 15445, 15580, 17640, 17715, 17730, 17895
USA KAIJ	5755, 9480
USA WBCQ	5110-LSB, 7415, 9330-LSB, 18910-LSB
USA WEWN	5810, 5850, 7560, 7570, 9450, 9955, 9975, 15785, 17595
USA WRMI	7385, 9955
USA WWCR	3215, 5070, 7465, 9985, 12160, 15825

Notes: Not all stations broadcast 24 Hours/Day. You may tune into a station and hear no broadcast.

LSB = Lower sideband / **USB** = Upper sideband / **V** = Frequency Varies

A larger list of shortwave radio stations can be found in our frequency appendixes, but a full listing is far beyond the scope of this book. In fact, books and directories are produced with hundreds of pages with channel frequency listings for radio shows which can be heard daily around the clock. We can however direct you to a group of web directories and some books that you can use for channel references and help in getting started into shortwave listening. During a power or communication outage, having a directory or list of channels to tune into will be important, however even without a directory or printed list you can still scan the shortwave band relatively easily for news.

Following are some popular sources for shortwave channels:

- http://www.shortwave.be/
- https://www.primetimeshortwave.com/
- http://swling.com/FindingStations.htm
- http://www.ccrane.com/shortwave-frequency-list
- http://hfradio.org/swbc/
- http://www.earthsignals.com/swfreqs.html
- http://www.hamuniverse.com/
- http://www.ccrane.com/shortwave-frequency-list
- http://swling.com/

Some good book references available include:

- International Shortwave Broadcast Guide – eBook by Gayle Van Horn
- The Worldwide Listening Guide by John Figliozzi
- Passport to World Band Radio by Lawrence MagneSh

In the International Shortwave Broadcast Guide Gayla Van Horn breaks down available radio stations by time of the day giving the listener literally thousands of radio channels to tune into for a variety of news, music and entertainment. As you can tune in to channels from around the world, you will also can get a perspective of news and life from across the world. Shortwave can be a fascinating window to the world.

4 - Ham Radio Repeater Directories

Having your radio is only the first step in successful radio communications. Once you have your radio you should seek out local repeaters in your area and tune in to repeater networks if you have some in your area. Many of the networks will operate network tests where radio hams contact the Network Control operator and identify themselves, pass messages, and just talk getting to know each other. These networks can be great sources for meeting other Hams in your area and getting a lay of the local land. Just by listening in on these networks you will get an understanding of the range that you have available through these repeater networks to speak to others in your area. In my own area of Tampa, I often speak to other hams in to 2-Meter/70-cm bands that are as much as a hundred miles from me. Distances in linked repeater networks are limited only by the size and number of the repeaters in the networks. In fact digital repeater networks linked together via computer links can communicate over vast distances.

There are several sources you can go to for a list of repeater directories. I recommend a combination of printed and online resources ranging from book, downloadable PDF printable lists, online repeater guides, and mobile phone Apps.

Printed Repeater Directories

Each year the ARRL produces a printed book in several forms from a desk reference down to a pocket-sized reference book. When I travel, I carry my pocket-sized edition with me so that if I need it and I cannot get on the Internet I have a handy resource in my backpack.

https://www.arrl.org/shop/The-ARRL-Repeater-Directory/

In addition to a repeater directory in book format, you can also visit the state or regional repeater council web site for your area. Often, they will have a printable directory available for your reference of repeaters in your area. See our section in appendix AG for a list of national repeater councils that will cover your area of the country. .

On your Phone or Tablet

In today's online connected world there are some great apps that you can install for your mobile phone or tablet. One great feature of these apps is that they use your phone's GPS to identify your location and then bring up the repeaters which are geographically closest to you. Some of these telephone apps include RFinder, Repeater Directory, and Repeater Book. These apps allow you to find repeaters that are geographically close to your location. Depending upon the height of the repeater, you can expect to be able to reach a repeater

at a distance from a few miles to up to 50 miles or even more with a 5-watt dual-band handheld radio.

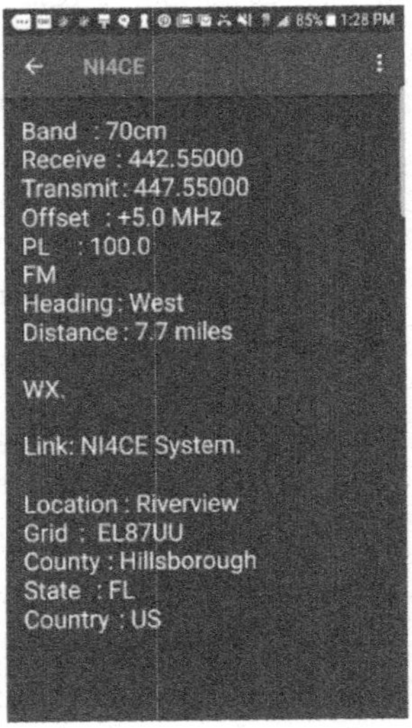

- RFinder – A very inexpensive subscription based app (About $10/Year) that lists repeaters based on distance from your location.
- Repeater Directory - The ARRL Repeater Directory. A Subscription based service with a free trial and includes maps for repeater locations near you.
- RepeaterBook – A Free repeater directory for your smart phone with distances from your current location to the repeater.

5 - Repeater directories on the Internet

For the most current information about repeaters check out the following websites. Most will provide detailed information about radio repeaters based upon types of repeaters, linked networks, or regional information.

- https://repeaterbook.com/
- http://www.arrl.org/repeaters
- http://www.rfinder.net/
- https://www.repeaterbook.com/
- http://www.radioreference.com/

Record, Write Down, Print and Document on Paper

For any frequency or repeater that you find that you wish to use from a source such as your state or region's own repeater council, radio club, or other resource it is important to record the specific settings required by that radio or repeater to connect with in a written form if you do not maintain hot and pre-programmed radios. You may need to refer to your paper notes if your power and/or internet connection is down to program your radios. And while at it make sure you have a printed copy of your radio's manual for manual programming reference. Settings which you need to record include:

Channel Name: Provide a reference name for the channel you are recording.

Frequency: Add the Radio Frequency

Tone Mode: Sets the mode used for transmitting or receiving squelch tones. This can be set to None, TSQL, DCTS or Cross

Tone: The CTCSS tone if tone mode is "Tone". If Tone mode is "Cross" then the values could be Tone, DTCS or blank.

Tone SQL: Sets CTCSS tone if Tone mode is set to "TSQL". If Tone mode is "Cross" then the values a value or blank.

Cross Mode: Controls the squelch behavior of the channel when the **Tone Mode** is set to **Cross.**

DTCS Code: Sets the DTCS Code to send if the **Tone Mode** is set to **DTCS**

RX DTCS Code: Sets the DTCS code to be used for receiver squelch

DTCS Pol: Sets the DTCS polarity of the transmitted code and the code used for the receive squelch any time DTCS is used for transmit or receive squelch.

Channel Name:	
Frequency:	
Tone Mode:	
Tone:	
Tone SQL:	
Cross Mode:	
DTCS Code:	
RX DTCS Code:	
DTCS Pol:	
Duplex:	
Offset:	
Mode:	

Offset: Sets difference amount for transmit and receive frequency if Duplex is "+" or is "-"

Mode: Sets mode of channel to FM, NFM, WFM, AM or DV.

Following is an example of a written record of several repeaters I communicate with in a linked repeater network in West Central Florida. Using this information, and the manual programming instructions, I can program my hand radios from scratch if necessary. The following chart can be found on multiple pages at the end of this book for convenience as well as pages of the above frequency form. You can use this book to log your frequencies of importance, copy the pages, or tear the pages out and use them. Whichever, try to write down your frequencies.

Repeater Name/Notes	Frequency	Offset	Tone Mode	Tone	Tone SQL	Modulation	Digi Code	Tune Step
NI4CE Linked Repeater Network – Riverview	442.5500	+ 5 MHz	Tone	100.00	88.5	FM		12.5 KHz
NI4CE Linked Repeater Network – Bartow	442.825	+5 MHz		100.00		FM		
NI4CE Linked Repeater Network – Verna UHF	442.9500	600 KHz		100.00		FM		5.0 KHz
NI4CE Linked Repeater Network – Verna VHF	145.430	+5 MHz		100.00		FM		
NI4CE Linked Repeater Network – Lake Placid	443.950	+5 MHz		100.00		FM		
NI4CE Linked Repeater Network – Holiday	443.450	+5 MHz		100.00		FM		

Advanced Topic: Setting up your own Repeater

After you start getting on the air with your radio you may get the opportunity to speak on some open repeaters in your area. This is a great opportunity to meet other hams in your area.

If you start wondering about how these repeaters work and the cost, let me fill you in a little. Repeaters are very expensive systems where the cost of standing up a repeater will run into thousands of dollars to cover cost of hardware and for the location to operate. Repeaters are designed to run dual frequencies (Duplex) with one channel for inbound communication and the other for outbound. They are also designed for high duty cycle operation (Lots of Transmit/Receive time).

But if you have an interest in playing with a repeater you can set up a small simplex record-and-forward repeater for under a hundred dollars. You will need a radio capable of DTMF and it should be licensed for the frequency you are using it on if you are going to use it for Ham, GMRS or MURS. The radio has to be legal for broadcasting on the frequency you are going to use. You can find some of these on Amazon or by googling "Simplex Repeater", but be sure to check the reviews. There are some good models and some not-so-good models. I have used Baofeng radios to set up inexpensive radios to learn with including a UV-82HP and a GMRS-V1 radio, both of which support 5 watts output and external antenna connections.

Emergency Service Repeater Networks

There are several Emergency Service repeater networks around the country which you can tap into for information during a natural emergency or disaster. Among these include:

- ARES http://www.arrl.org/ares
- RACES http://www.usraces.org/
- Weather Nets https://www.weather.gov/pbz/hamradio
- SKYWARN https://www.skywarn.org/

For a listing of these in your area visit RepeaterBook.com, North American Repeaters, and select your state. Many of these networks will operate on existing repeater networks and run meetings or tests at different times of the days or weeks. These networks consist of volunteers and members of many of your local Emergency Operations Centers who participate to train volunteers and testing of equipment for use during emergency situations. In my own area of Florida I have often "Checked In" to the networks which essentially allows the control operator to log who you are and your area as well as allowing you to test reception of your radio and participate.

If an emergency occurs such as inclement weather or natural disaster (In Florida we have Hurricanes and strong summer storms such as the 2017 IRMA Hurricane). During hurricane Irma Hams were able to both listen and, if necessary report, weather conditions being observed. While many Floridians felt the effects of wide spread power outages affecting cellular phone coverage, Hams on as little as 5-watt hand held radios and mobile radios were still able to participate and report problems. For days after the storm we had regular participants checking in on the networks still in affected areas.

6 –Repeater Councils and Clubs

Most states or geographical areas have independent repeater councils. These councils are responsible for working out radio Band plans for their areas and most will also publish a state or area directory of repeater systems and networks. From the Internet, you can search for these by using as a search term "Repeater Council *YOUR-STATE"* and go to the website you find.

In Appendix AG near the end of this book you will find a list of the various state and regional repeater councils. Each of these councils holds the responsibility for coordinating the frequencies in their areas for repeaters within their areas of responsibility. These councils also established an agreed upon band plan for the areas they oversee. These band plans are used by radio operators to avoid creating interference on repeater networks and to coordinate good radio use.

Most of the websites for the repeater councils will include the band-plans for your state or region which have been agreed upon and established for the areas covered by that council. These band plans will cover all HAM bands including popular 70-cm and 2-meter plans used by inexpensive hand-held radios and bands in all other amateur frequencies available to licensed ham radio holders (Frequencies available are always based upon license level held).

Radio Clubs

Local radio clubs and an excellent place to meet other Hams and to find experienced ham radio operators to learn from. Many of these clubs share the expense and run their own repeaters covering their areas. Local clubs can be found through the ARRL Website at www.arrl.org and searching for "Clubs". Members of the clubs will welcome new hams and will be a great local source of information for what networks and frequencies are popular in your area. Not that other than national standard calling frequencies, the frequencies used by Ham repeaters are not standardized other than being within the range of frequencies established for the band and agreed upon by the Repeater councils for that area.

Many radio clubs hold regularly scheduled testing sessions for anyone wishing to get their first license (Technician's class) or for existing hams who are upgrading to their General or Extra class licenses. Testing is affordable (in 2017/2018 only $15.00 from AARL VECs, less from some others) and licensing is required for operating on any amateur frequencies. For testing schedules and locations you can check out the ARRL.org website.

7 - Antenna Systems

In radio, there are two factors that determine the quality of the signal from transmitter to receiver: Power output and Antenna signal.

Power out is the brute-force ability to push the signal out of the radio into the airwaves to be received by a receiving station at some location. For a very large part the power of the radio adds more to the clarity of the signal being pushed out than to the distance of the signal. With a good antenna and proper line of site a 5-watt hand-held radio can communicate 50 miles or more based on conditions. The same radio communicating to another handheld radio in a densely wooded area or city area may, under worst-case conditions, have a problem communicating within a couple of miles.

The Antenna signal is also dependent upon the quality of the antenna and the radio manufacturer and is every bit as important as power. But the antenna signal is also dependent upon the design of the antenna and how it is matched up with the radio and for the frequencies it is being used for. A good radio with a good antenna the radio can function better with less power than a more powerful radio with a bad antenna. Where power is the muscle in radio, consider the antenna more in many ways the artist needing to perform not just with the radio, but also with the environment. A good radio with a good antenna will also fail quickly if used within a confined area with obstacles and in an environment, not conducive to the radio signal (Basement, Inside of a building or warehouse, etc).

The factors that determine the ability of the radio signal to be well transmitted and received include, but are not limited to the following:

- Radio frequency and properties of the frequency
- Power of the radio pushing out the signal
- Antenna design itself – Includes antenna resonance and performance
- Obstacles and interference (Buildings, natural surroundings such as trees, hills, mountains including Line-Of-Sight)
- Characteristics of the Radio Frequency (Propagation of the selected frequency, Ground travel, signal penetration, frequency used)

To understand how these factors play into our ability to communicate with another person at another location, let's review each of these factors individually.

Antenna Design

As mentioned before, a good antenna on a weak radio will outperform a poor antenna on a more powerful antenna. Let's look at GMRS radios as an excellent example of this.

Over the past several years' manufacturers have packaged GMRS/FRS radios into combined units and marketed these radios with outrageous specifications. "25 Miles", "36 Miles" or even "50 Miles" are marked on packages. While a good GMRS Radio *Could* communicate 50 miles with another radio (Under perfect conditions), the idea of these packaged combo FRS/GMRS radios doing that is a joke. I have a set of 5 watt GMRS/FRS radios and consider myself fortunate to get a mile. But that is ok for those because I purchased them for use hiking in a group and they are convenient for that purpose. I also have a good dedicated GMRS radio and though I have never spoken extremely long distances with it, I have spoken over 5 miles on it under good clear conditions. What is the difference between both radios that are both 5-watt radios? The dedicated GMRS Radio I own allows me to connect an external antenna to it pushing the range much farther than the combo-pack radios allow.

A Note about Combo FRS/GMRS Radios
Under 2017 GMRS Radio rules revised by the FCC, these radios are no longer to be considered for use or sale as GMRS Radios. Antenna Rules for Combo radios must follow the more restrictive rules for FRS radios and allow only fixed limited range antennas on the units. Combo radios putting out 5 watts of power likely will be restricted in the future to the 2-watt limit of FRS. 2 watts still works well as do these radios for short range communicating, but not as effectively as better GMRS dedicated radio. For an example of a good GMRS dedicated 5-watt radio allowing changeable antennas check out the Baofeng GMRS-V1 radio.

The FRS/GMRS radio I have is limited because under the FCC Radio rules for FRS, the design of the antenna is that it must be a permanently fixed antenna to the radio and cannot be changed out. On my dedicated GMRS Radio with the same power, the antenna for that radio is removable and I was communicating using a modified external antenna to another contact hitting a repeater located ten miles away that was elevated 30 feet up. (20 Feet is the legal limit for GMRS antenna height *above* the structure it is mounted on which was approximately 10 (or so) feet high. And we were communicating in a rural area in Florida (Flat Florida) with no buildings and no dense growth. So, two radios, same power, completely different antenna designs and completely different results.

Similarly, I often talk driving throughout the Tampa Bay area with my handheld 5-watt Ham radio using a mobile antenna on my vehicle to other hams bouncing off a repeater in our area mounted 800 feet up on a significant shared antenna tower. The distances I drive in the area are roughly 20-35 miles from that antenna.

Antenna Resonance and Tuning

Antennas need to be tuned to proper lengths based upon the wavelength of the radio frequency to maximize the power of the transmitter. A badly matched antenna will result in a loss of power and signal coming out from the transmitter affecting radio range. For mobile and handheld radios, most radios use a fractional wave antenna at half or quarter measure of the wavelength which, though not the full wavelength of the radio wave, still allows for an efficient length for the radio signal. The lengths used are relative to the length of the radio wave. To understand what this length is you need to know the formula for determining the wavelength which is:

Length of radio wave (Meters) = Speed of radio wave (Meters) / Frequency in MHz

In the chart on page 24 are some examples of wavelengths in meters, then shown in feet for the full wavelength, half length, and quarter wavelength in Feet and inches based on various frequencies.

As you can see in this chart, the lower the frequency such as for CB Radio, the longer the wavelength. This also explains why handheld CB radios are not nearly as common as 2-meter or 70-centimeter radios. Can you see how inconvenient walking around with a 9-foot quarter wave CB antenna would be versus one that is less than 2 feet for VHF, about 6.6 inches for a UHF Radio, 1.6 inches for a cell phone or under 1.3 inches for XM Radio? (Ever wonder about that little XM Nub on the top of your car?) This means that hand-held CB radios use a much less efficient antenna which is shorter for convenience. This convenience costs the radio range and effectiveness.

Back to the FRS/GMRS radio example from earlier. My bubble pack GMRS Radio has about a 3-inch antenna on it. My dedicated GMRS radio connects via Coax cable feed line to a half-wave GMRS external antenna mounted *outside* of my car. Which would you think performs better under the same power? The GMRS naturally.

Obstacles and Interference

How much *Stuff* is between you and who you are speaking to?

If you are in a building, or in a city where you must talk between a lot of stuff, and a lot of dense stuff (Metal, Concrete, buildings, stadiums, cars and trucks) you are going to have a much more difficult time talking than if you are in a rural area, on the water in boats, or in the air (Aircraft). Airband radios can communicate much greater ranges than any other types not because of the radios, power or antennas, but because they have nothing between the aircraft and either the other aircraft or the radio tower. Astronauts on board the International Space Station commonly communicate with the ground using just a few watts of power at distances hundreds or even a thousand miles.

Your environment is the single most critical factor in your ability to communicate. Depending upon the frequencies you are communicating on, some are more vulnerable to obstacles and the ability to travel than others, but all are susceptible to the interference caused by obstacles.

Chart: Radio Wave lengths for various frequencies

Description of Signal	Frequency (Hertz)	Wavelength (Meters)	Wavelength (Feet)	Half Wave (Feet)	Quarter Wave (Feet)	Quarter Wave (Inches)
CB Radio	27,000,000	11.11	36.44	18.22	9.11	109.33
VHF 2M Ham Call	146,520,000	2.05	6.72	3.36	1.68	20.15
70cm Ham Call	446,000,000	0.67	2.21	1.10	0.55	6.62
33cm Ham Call	927,500,000	0.32	1.06	0.53	0.27	3.18
GMRS Channel	462,550,000	0.65	2.13	1.06	0.53	6.38
Marine Radio Calling	156,450,000	1.92	6.29	3.14	1.57	18.87
MURS Channel 1	151,820,000	1.98	6.48	3.24	1.62	19.44
GSM Cell Phone	1,850,000,000	0.16	0.53	0.27	0.13	1.60
XM Radio	2,332,500,000	0.13	0.42	0.21	0.11	1.27

In this chart, the longest wavelength is that of CB Radios at over 36 feet, over 9 feet for a quarter wave antenna. VHF/UHF Radios which are popular in handset formats have more convenient size antennas of just a few inches for a UHF quarter-wave antenna (The common "Rubber Duck" antenna on a hand-held radio).

Line of Sight Radio and Antenna Height

Most radio operates at a level of "Line of Sight" meaning direct distance from antenna to antenna. Depending upon frequency you are working in, some frequencies are more rigid to line of sight than others, but all follow this rule with some twists depending on frequency. As your frequency increases, such as into the VHF and UHF Frequencies, this rule becomes more rigid.

What we are referring to by line of sight is how the radio signal transmits from point to point taking the curvature of the earth into account. Though trees, buildings and other natural as well as man-made obstacles to make a difference, generally we are here referring to the curvature of the earth and significant obstacles such as mountains,

So, what is our line of site from a handheld radio?

Take a person standing in the open holding a handheld radio (Position C in our diagram). The normal curvature of the earth dictates that the line of sight with the antenna raised at five feet above the ground will be approximately 3 miles (2.99 miles from position C to position A) from that point to the farthest point where the earth curves. Beyond that point falls below the curvature of the earth. If a second person is standing 3 miles beyond that point with his antenna 6 feet above the curvature of the earth, then together their antennas are seeing each other at 6 miles (Position C to position C). After that signal loss becomes great.

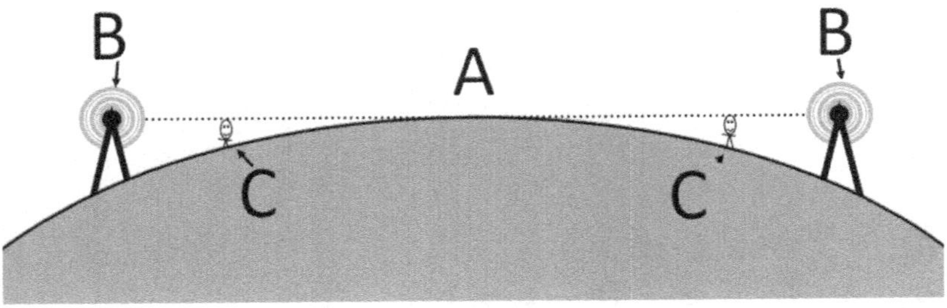

$$horizon_{(km)} = 3.569 \times \sqrt{height_{(meters)}}$$

So, is 6 miles the range limit? Well, of course not. If you are talking to another radio operator from radio to radio, often referred to as simplex radio operation (Both users transmit and receive on the SAME frequency without a frequency offset) then, well, yes, 6 miles would be the limit. Increasing the height of either antenna will extend this distance either by directly raising the antenna or by standing at a higher level (Balcony of a building, on top of a building or tower, on a hill, or on a mountain) then the range will increase. Radio operators talking from high points, such as mountain to mountain with large distances between them but no obstacles (50 miles for instance) can communicate on low wattage radios (5 watts for example) quite well. Again though, this depends on the height of your antenna and obstacles in between both radios. Talking to another radio operator at 50 miles from the peaks of two mountains or high hills with direct line of sight communicating is common.

In our example on the previous page, we have two tall antennas which each have a height of 25 feet. If I am communicating to that station and my antenna is 5 feet high then our combined range we can communicate from is 10 miles (25 Feet gives a distance to the radio horizon of 7 miles plus my range of 3 miles at 5 feet high).

For more on line of sight radio distance calculations visit HamUniverse.com at http://www.hamuniverse.com/lineofsightcalculator.html .

Enter Repeater Systems.

In radio, repeaters can be used to extend the range of your radio communication. A radio operator standing with an antenna height of 6 feet talking to a repeater with an antenna on a tower that is 500' in the air (A Commercial tower) can talk at a range of 35 miles. In my own area of Florida I communicate throughout much of the Tampa Bay area through a repeater on an antenna at 860 feet in the air from over 40 miles away. And since that repeater is part of a 5-repeater network, I frequently talk to other radio operators over 100 miles away!

Characteristics of Radio Frequencies

Lastly, we need to discuss and understand the differing qualities of radio signals communicating at differing frequencies. We just spoke about "Line of Sight" where radio signals like to travel in straight lines. But radio signals at lower frequencies have more of an ability to bounce or reflect around the curvature of the earth or bounce from the atmosphere than do higher frequency radio signals.

The ability of a radio wavelength to bounce or travel around the curvature of the earth or through the atmosphere is called Propagation. Propagation is what allows an AM radio signal to travel hundreds of miles or allows a HAM radio operator to communicate across the country or across continents, even to travel part way around the world. Many factors play into propagation such as solar activity and weather conditions. This allows a radio operator one day or one minute to reach several thousands of miles away, and then right after loose that contact and not be able to reach them again.

Propagation is common with low-frequency radio waves which have longer wavelengths. In fact, it is a characteristic of these lower frequency. This leads to the rule by the FCC for AM Broadcast stations to have to reduce their power at night after the sun goes down and the atmosphere is more conducive to radio signal propagation traveling further. Otherwise this would cause bleed-over effects between radio stations at night if power were not reduced.

Propagation is what allows shortwave broadcasters to transmit extreme distances to other countries. This is how stations such as "Voice of America" are able to transmit to foreign countries who are not allowed to hear our news, and why other countries including Russia, China, North Korea and Cuba have radio broadcasts targeting to the United States.

As the frequencies increase and the radio wavelengths shorten, the ability to bounce reduces and radio waves better penetrate structures. Thus, our cell phone signals work better indoors than shortwave signals. In radio communication, this is also why VHF does slightly better for communicating outdoors and UHF tends to slightly better indoors.

An excellent analogy I once heard is as follows. Imagine walking through a building holding a stick that is 6.72 feet long and holding it without turning it. This is the length of a 2-meter VHF wavelength. You keep bumping into things and hitting things. Now walk with a stick that is just 2.21 feet long – the length of a 70-cm UHF wavelength. It is much easier to move throughout the building with the short stick than the long stick.

UHF signals (Short stick) penetrate buildings better than VHF frequencies (Long stick). So, while this sounds better, the reverse is true of a signal's ability to propagate outside. A VHF signal generally travels farther outdoors where there are few obstacles. And likewise, by dropping into the HF bands with even lower frequencies, our distance further increases. This allows 6-meter and 10-meter radios to communicate farther outdoors without repeaters.

8 - Selecting a Radio – Quality and Features

As a Ham, I cut my teeth learning ham radio and earning my Technician's license as an Amateur radio geek using a $30 Baofeng handheld radio. It is a good little radio and has been a fun radio to use. Today that radio has been retired to my emergency radio kit along with my emergency flashlights, lanterns, rotated water and emergency food in favor of a slightly more expensive Baofeng and slightly better model that I carry with me as well as a nice model mobile radio I keep in my vehicle. I grew up in Florida and worked on a military base in 2004 when we planned for evacuations four times in September that year, so I take hurricane planning serious. It is also a reason I dug into Ham radio and emergency communications.

This chapter has been born from my reading radio reviews and watching videos from experts teaching various topics from radio frequencies, to propagations, to repeater education. I find that 99 percent of these videos are useful, but every now and then I hit a video from an "Expert" who gets into radio quality. One expert I recall was educating people that the minimum "Must Have" radio was going to start at $200 each and anything less was simply junk and could not be relied upon. Another expert I also watched recently was discussing the cheesy gimmicks that the cheap radios have sometimes including a flashlight and an FM radio. Well, I use my FM Radio in my old pickup that I drive on remote hikes (I don't care about it getting scratched up in the brush) and that flashlight has come in useful on more than a few occasions while out hiking. That same reviewer proceeded to show his minimum recommendations for radios, all with a retail price of $500 or more. Well I whole heartedly disagree.

I do know that the cheaper radios have their deficiencies and naturally an expensive radio will have a lot of benefits over an inexpensive model. So,000 I want to talk about some limitations and some benefits of going "Bigger". I likely won't be putting out for a $600 handheld without a substantiation benefit any time soon. Here are some of the features I want to mention that you may want to consider and if they are important for you.

- Durability
  I don't rank the inexpensive consumer radios high on this scale. You get what you pay for – but I also don't confuse dependability with

durability. My $50 Baofeng (Upgraded from my $30 Baofeng) has been great for months for hiking trips, in the car, or chatting from my couch. But if I drop it – I would not be surprised if it stops working. It is dependable, not as much durable. But still a great little radio.

- Waterproof
  My $50 radio is waterproof. Per the manufacturer I can drop it in water and it will still work. Leave it in for a long period? I have not tested it, but I would not be afraid of being caught out hiking in the rain with it. My first little UV-5R? As I understand it that radio will fry in the rain. It is what it is. But I would look at some level of water proof or resistance as important when you evaluate a radio. Many manufacturers offer water resistant and water proof radios with an "IPX" rating level. IPX4 would be the minimum I would recommend for getting caught in the rain up to IPX8 allowing or immersion beyond 1 Meter.

- Flashlight
  Gimmick – but I have used mine while hiking. Usually because it is handy, not because it is my only light. But even a gimmicky light can be useful and handy sometimes.

- FM Radio
  I consider also useful and use mine regularly. The radio on the Baofeng works great sometimes while hiking and automatically silences in favor or an incoming message on one of up to two stations I am monitoring.

- GPS
  Don't have it on my radio, but consider this a great feature if hiking or remote with others and staying in contact. Some radios will transmit your GPS location out when communicating – A Great feature for use in emergencies.

- Digital Radio
  The new horizon of radio. Digital offers cleaner signals over similar distances, though the range usually cuts shorter than the max range of an analog signal. You will also need to be on a digital network. Digital radio will also open a new range of features including talk groups, filtered communication, location data, and more.

- Text Messaging
  The ability to send short messages from radio to radio. Not a feature of the inexpensive radios.

So, what type of radio would be a good starter radio?

This depends on a few factors. If you are living in an area where there is a repeater network for the 70cm band (440MHz) or the 2 Meter band (146MHz) and there is activity on the repeaters, then you could start out with a simple hand held. Performance on a hand-held is going to be depending on the antenna you have, distance from the repeaters, and surrounding geography factors such as buildings, mountains, and hills. In my area of Florida on the west coast (Flat) I have access to a larger repeater network. I have spoken to others though in more remote areas where a simple hand-held radio is useless. Remember range on these radios is short and line of sight.

But if this type of radio does work for you, you can pick up a radio such as a Baofeng for as little as $30. This gives you an inexpensive radio to start with and learn with. Better brands from manufacturers such as Kenwood, Icon, and Yaesu offer more advanced features at a higher cost and higher quality radios. Depending on features these better radios can run into hundreds of dollars and should be considered based on the features you need. But just to get started, a $30 radio will get you on the air.

Keep in mind one of the factors is range of these radios. If you can hit a repeater network with a good quality signal then you will have a connection to other hams. If your area does not benefit from such a network, then your range with one of these handheld radios will limit you so consider a mobile radio with better range through an external antenna and more power. You can still step into a more powerful mobile radio for around $100 from several vendors, plus cost of an antenna. Even though these are "Mobile" they can run from a 12-volt power supply and be run from your house. For the addition of an external antenna, cable and power supply figure in another $100 or so.

With your first license, a Technician's class, you will have access to communicate on 70-cm, 2-meter, and very limited 1.25-meter bands with voice. If you want or need to talk further without a repeater network then you will need to step into a radio capable of 6-Meter (VHF), 10-Meter (HF) or higher bands. There are several mobile radios which will offer 6 and 10 meter communications starting out in the $200-$300 price range and moving up. Though more expensive, with these radios you are getting into greater distance through propagation of radio signals (Higher the band the better). To operate on these bands you will need to obtain your General Radio license which opens up most of the bands to you including the higher bands up to 160-meters.

Stepping into better radios will offer you better capabilities for distance, but here is also where features and quality of the radios will grow as well as costs which can quickly break over $1,000 for just the radio. Add power supplies, enhanced microphones, antennas and other gear, Ham radio quickly becomes an investment. But depending on your location and need many hours of enjoyment can come even from the simplest of radios and the lowest of investments. Determine what you want to do with the radio to determine what the best option is for you.

For short-range use, hiking, talking on repeaters, and emergency use consider starting with a simple hand-held. Technician's License needed.

For participating in emergency groups such as Skywarn, Ares, RACES or local Emergency Operation groups, a simple hand-held may also be sufficient. During hurricane Irma here in Florida I was able to monitor emergency notifications and reporting by the local Skywarn network with news coming in throughout central Florida from a $50 hand-held radio. Technician's License needed.

For distance communicating you need to get into 10-meter and higher bands. General License needed. Decide what type of communicating you want to do, the channels and frequencies. Pick up an inexpensive 70cm/2m radio and learn some ropes.

9 - Advanced Radio Topics

Digital Radio, Trunked Radio Systems, and Digital Encryption

The original goal of this book was to provide written reference information for national radio channels, frequency information, and some basic information about radio systems intended primarily for the radio novice and newcomers to Ham radio. My hope has been to close some of the information gaps and answer many questions that newcomers have when getting started. With the price drop of entry level radio systems there has been an increase in the number of people purchasing to learn radio technologies, and a lot of radio systems sold to those who fall into the "Prepper" mentality but have no clear understanding of the use or laws after purchasing.

As I hope many of your answers have been answered, we still have some areas which I want to touch briefly on so the reader is aware of these topics, however we will not go into any real depth on these. But they should be mentioned for the newcomers into radio to be aware of as you navigate into the technology. These topics may affect your ability to receive and tap into some radio systems.

Digital Radio

Radio has been around since 1895 when it was invented by Italian inventor Guglielmo Marconi and since then it has been ever changing, In the past 20 or so years a new flavor – digital radio, has been making headway into the hands of ham operators as well as breaking into government and commercial markets. Digital offers many new advances including the ability to transmit digital data with your communication, location and movement information, more efficiency of existing radio channels and abilities to share channels for more conversations using the same frequency bandwidth

Analog is still the popular standard for Ham radio across the world. But new capabilities that are coming are helping to push the interest in digital radio systems. As the airwaves continue to crowd digital offers new ways for getting more conversations and information onto those same frequencies. Organizations, businesses and government are turning to digital to solve these new communication challenges.

Privacy challenges are being solved through encryption that can be added through digital solutions. Encryption allows groups or organizations to define privacy keys for their conversations which will block outsiders from listening in. For those interested in radio for communication within your own groups this will not be a concern. But for anyone interested in monitoring, it may or may not affect you depending on what your local organizations are doing.

For the benefits that Digital is promising, there are still differences among standards which keeps each system from being open. P25, DMR, NXDN and other standards offer differing implementations keeping them from being compatible with each other. When choosing digital you must select a standard to follow, and if needing to communicate across standards, well this may require multiple radios. Prices are significantly higher for digital radio which will keep the cost of a good radio into the hundreds of dollars.

Keep in mind – All digital solutions were designed for use in commercial, government and emergency environments. All implementations bring great new advantages for use as a system approach and use for amateur ham use was not a goal of any of these systems. For the amateur, analog is still the best solution for learning and cutting your teeth on. Manufacturers such as Baofeng, TYT and others who are now breaking into DMR are each having challenges with their own implementations. If you do decide to venture into the world of Digital radio be sure to check reviews of the models before you purchase. The only sub-$100 DMR radio at the time of this writing is marketed as a DMR radio and in fact is, however has implementation problems keeping it from working well with other DMR radios and repeater networks making it a poor option as a fully compatible radio. Check radio reviews and particularly watch the reviews on YouTube.

At the time of this writing some low-priced digital radios have been coming to market breaking below the $100 barrier. Be aware however that some of these implementations are not compatible with repeater networks or do not implement the digital protocols in a way that they are completely compatible with other radio systems. For hams, be very careful when making your selections. Read the reviews and follow YouTube reviews on the radios.

Trunked Radio Systems

Another topic for discussion to understand is that of Trunked Radio systems. These systems are implemented in many business, commercial and government radio networks as a method to allow a fixed number of channels to be more efficiently used by a larger number of users. Through Trunking, a set of radio channels are grouped together and users can share these frequencies in a way that they can be automatically provided a frequency for communication based on availability. The trunked system assigns a user to a frequency that is open for their communication. The next time the user communicates they may be coming in through the same frequency or another.

Trunked Telephone Lines shared between cities to maximize efficiency

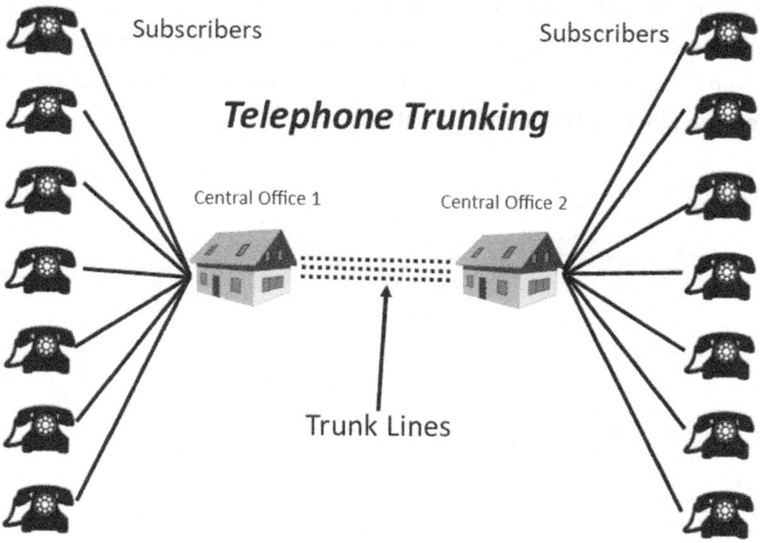

Trunked radio works in a similar way that the Telephone companies have used trunked telephone systems for decades – Many users share a small number of frequencies as needed based on the assumption that not everyone will need a dedicated frequency at the same time.

If you are scanning and listening in on a frequency used by a trunked radio system you may hear parts of conversations but not the entire conversation. Trunking allows the fixed frequencies to be used much more efficiently over time. Some models of radio scanners will allow proper monitoring of some trunked networks.

Radio Encryption

Radio encryption is being used more now by businesses, communities and government to keep communication private. Encryption can be accomplished on both analog and digital radio networks and there are many methods of accomplishing it. If listening to an encrypted communication it will sound like static or noise unintelligible. Decrypting an encrypted signal is illegal as is effecting the signal.

As more radio systems including government and public service systems convert to digital for better use of bandwidth and efficiency, many systems are also including encryption in their radio systems. This is important to realize as though many of the frequencies which are established by communities and organizations remain the same, encryption now being used on these radio systems will prevent third parties from being able to tune in and listen to the communications.

10 - Notes for bargain radio buyers

This section is important to clarify some points for the purchasers of many of the newer Chinese and other sourced ham radios now on the market. These are radios like the extremely popular Baofeng radios. We are not going to knock the radios – I own multiple myself and they are great basic radios and have allowed many new hams to get into the hobby very inexpensively. But here are a few very important things you need to know when purchasing these radios.

1. **Amateur HAM Radios are not out-of-the box legal for YOU to use.** This is referring to the popular Baofeng and similar radios widely available now. What this means – To use these radios on the HAM frequencies in the 70-cm and 2-meter amateur bands you MUST be licensed. Yes, they will work. But using them without a license is not legal. Yes, if something happens and things go crazy they will be great to have. But legal use requires a license.

2. **Even with a license, they are not legal for GMRS Frequency use.** Most are not legal as they have not been FCC certified for GMRS frequencies. Baofeng does have a legal GMRS-V1 radio which is a good radio. But most of the popular models are NOT legal.

3. **These are NOT legal for FRS Radio use.** Like #2 above – They have not been certified by the FCC for legal use on FRS frequencies. Also – These radios on LOW power still broadcast at 1 watt which is double the ½ watt allowable on most FRS frequencies and under the new 2-watt power limit for *some* FRS channels. They are still however *NOT* legal for use on FRS Frequencies.

4. **They are not legal for MURS, or Marine use**, or any other channel use outside of HAM (And for HAM you must have a FCC License).

5. **BROADCASTING ON THE WRONG FREQUENCY OR AT THE WRONG POWER LEVEL CAN RESULT IN FINES BY THE FCC IN THE THOUSANDS OF DOLLARS.**

6. **Marine and Airband Frequencies**
   These frequencies are strictly reserved for use by mariners or those working aircraft communication such as ground or vehicle pilots. Though licensing my not be specifically required for these radios operators, use in the United States for their designated purposes is considered legal. Use for non-designated purposes is Illegal (Don't use Marine channels, hiking, hunting, or for car-to-car communication – This is NOT LEGAL).

To be legal for use, any radio must have the following:

1. An FCC Sticker on the back of the radio (Usually under the battery) indicating what they are legal for.
2. The radio must be legal for the specific use (It could be for HAM but not GMRS. Or it could be for GMRS but not FRS, MURS or other bands).
3. If in using an illegal radio you cause interference and are caught you could pay fines of up to several thousands of dollars and lose your license if you hold a radio license.

Many of these inexpensive radios are great bargains and will help to get you into the hobby of HAM radio just fine. I carry an inexpensive Baofeng myself. And without a license you can still legally listen and monitor channels on these radios. In fact I would encourage you to – Learn how to use them and where the Nets are in your area. It is in the broadcast side where you could cause harmful interference with other radio operators, homeowners with cell phones, televisions, or other electronics receiving your signals. (In reality, a low-power transmitter won't likely cause interference, but let's stay legal anyway.)

Some brands of radios do have the ability to turn off the broadcast capability on programmed channels to keep you from accidentally keying up the microphone and creating interference. Many do not however, though you can program the power to the radio to a minimal level (1 Watt usually) so that if you do key the microphone you won't as likely be heard.

Radio without a Radio – EchoLink

A type of Ham radio which I want to briefly touch on is actually not a radio system, but a computer radio application called EchoLink. This is an application can be installed on your home computer or on a smart phone device and uses computer server/networking to connect to other users around the globe. All users MUST be licensed Ham radio operators. The application supports remote nodes connected to radio networks or connecting your own radio to your computer for remote use.

What EchoLink allows is the ability to travel and still connect to operators who are using the system back to your own home areas or networks. In describing the system, I listened to one presenter describe this as "Augmented Radio". It is not true ham radio and will not operate where most hams will be able to during an emergency or disaster. But still – what a wonderful way to use your skills as a Ham radio operator to talk to users around the world without the challenge of propagation, antennas and DX challenge – though these challenges are what make Ham radio a true draw to many radio hams.

The system currently boasts of over 200,000 registered users across the world in 193 countries with about 6,000 users online at any given time. The software supports a variety of hardware interfaces which allow you to connect your computer to radios for remote communication through the internet to your radio at home and then to your home area.

If you are interested in EchoLink check out the website which can be found at http://www.echolink.org. Remember it requires confirmation of a valid Ham radio license to use but can offer a door beyond local radio without the expense of a bigger radio system. Advanced use of EchoLink opens the door to being able to stay in touch with your local nets while on the road and traveling through a laptop computer or a smart phone device.

11 - Emergency Radio Protocols

The image below is from RadioMaster Reports website (See link below) and lists frequencies that can be used during an emergency such as a national disaster for communication. These frequencies assume Simplex communication (Radio to Radio vs. Radio to Repeater to Radio) so radio range will vary depending upon your type of radio.

(Radio Masters Website: https://radiofreeq.wordpress.com/)

Not all radio frequencies will be available from all types of radios. Handheld, mobile and desktop HAM radios will be able to reach to many of these channels for FRS, GMRS, MURS, Marine and HAM use.

PREPPER & SURVIVALIST SHTF FREQUENCIES
2-WAY RADIO COMMUNICATIONS

RADIO SERVICE	CHANNEL NAME	FREQUENCY MHZ MODE	DESCRIPTION
FRS UHF	FRS 3	462.6125 FM	PREPPER
GMRS UHF	GMRS17	462.6000 FM	SURVIVALIST
GMRS UHF	GMRS20	462.675+ FM	PL141.3RPTR+5MHz
PMR UHF	PMR 3	446.03125FM	PREPPER
MURS VHF	MURS 3	151.940 FM	PREPPER
CB AM	CB 3AM	26.985 AM	PREPPER
CB AM	CB 9AM	27.065 AM	HIGHWAY SAFETY
CB SSB	CB 36U	27.365 USB	SURVIVALIST
CB SSB	CB 37U	27.375 USB	PREPPER
CB FREEBAND	FB368U	27.368 USB	SURVIVALIST
CB FREEBAND	FB378U	27.378 USB	PREPPER
CB FREEBAND	FB425U	27.425 USB	SURVIVALIST
LOWBAND VHF	LOW334	33.400 FM	SURVIVALIST
HAM UHF	HAM U3	446.030 FM	PREPPER
HAM VHF	HAM 42	146.420 FM	PREPPER
HAM VHF	HAM 52	146.520 FM	HAM CALLING
HAM VHF	HAM 55	146.550 FM	SURVIVALIST
HAM HF	HAM10M	28.305 USB	PREPPER
HAM HF	HAM20M	14.242 USB	PREPPER
HAM HF	HAM40M	7.242 LSB	PREPPER NETS
HAM HF	HAM60M	5.357 USB	SURVIVALIST NVIS
HAM HF	HAM80M	3.818 LSB	PREPPER NETS
LAND SAR VHF	SAREMT	155.160 FM	SEARCH&RESCUE
MARINE VHF	MAR 16	156.800 FM	SAFETY CALLING
MARINE VHF	MAR 72	156.625 FM	BOAT PREPPER

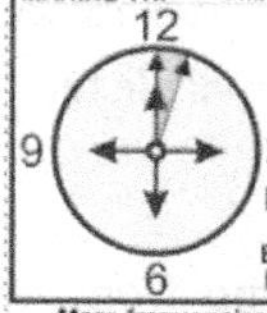

3-3-3
RADIO PLAN
CHANNEL 3
EVERY 3 HOURS
FOR 3 MINUTES

2013 Public Domain
Source:
RADIOMASTER
REPORTS

VERSION: SHTF FREQ LIST 2013E

More frequencies at: RADIOFREEQ.WORDPRESS.COM

An emergency radio plan is a plan that you have for communications during an emergency or an event. This is an overall plan that includes being able to stay on top of local or national news that is occurring through portable radios, or if necessary being able to "Plug In" to radio communications from longer distances through Shortwave radio. A good radio plan also includes the use of two-way radios to be able to stay in touch with family members and other contacts nearby or at a distance.

All radio plans start out with a family communication plan. A plan that you put into motion to bring your family together in the event of an emergency or an event. Your family plan and radio plan will both be a part of an overall preparation plan to enable you and your family to get through an event together and with as little risk as possible.

Why do you need these plans? Well, let's look at a scenario.

Ok – So the situation is that for some reason, either from a storm, to an earthquake, sudden power loss, or some other reason, all communication has stopped. This can occur for many reasons which we will not cover but the result is the same – quiet telephones and the need to get your family together for safety. If regular telephone communication fails for any reason the public will immediately jump onto their cell phones to communicate. During the attack of New York City in September of 2011, cellular systems were almost immediately overwhelmed keeping the public from being able to use their phones.

During a power failure, though Cellular systems will immediately fail over to backup power which is limited and cellular towers will begin to drop from the network as early as just two hours after power loss. In an extended power loss, this will make normal communications impossible.

So, your emergency radio plan with your family first must start with an emergency communications plan. In my family, I have worked it out so that my kids will immediately go to a close friend's home who is also a nearby neighbor and in walking distance. They also have been taught to carry a spare portable radio which I keep for our hiking and backpacking trips. This is a legal and professional level radio which has an external antenna and a longer range than the normal blister-pack GMRS Radios which can be found in your local retailer.

For myself, if travel home from my office which is 30 miles away is impossible, then I will reach out to her through my own portable HAM Radio which is programmed to the same frequencies and to a local repeater network in my area which is itself tied in with an emergency network. I will be able to reach the kids and family much sooner than I can get there and plan our reunion for safety. It needs to be noted though – If the repeater network fails, then my radio range will be roughly 5 miles. By having a schedule in a radio plan of when we will try to communicate we can conserve our batteries.

3-3-3 Radio Plan

The 3-3-3 Survivalist radio plan works with most any type of 2-way radio types available and dictates both stations as well as timing and how to use the radio. The purpose of the plan is to provide a schedule through which you can keep your portable radio turned off to conserve the battery power, but have a schedule that is known through which you can turn on the radio and tune into specific channels to reach others. This is a plan that is published in the prepper and survivalist community and following it during an emergency event may help you to reach someone.

The plan works for all types of two-way radios including FRS, GMRS, MURS, CB, Marine or HAM. For all channels, other than HAM there are designated channel numbers, of which Channel 3 is identified in this plan for making contact. For HAM radio operators, each radio band (2-Meter, 70-cm, etc.) has a "Simplex Calling Frequency". What this refers to by Simplex is that you will be using your radio on a frequency not used by a repeater, but rather for radio-to-radio communicating.

Keep in mind that on a simplex frequency, distance is more dependent upon power and antenna for your radios. For this reason, most of your smaller radios such as FRS/GMRS combo radios and MURS Radios (2 Watts) will have limited range and connectivity. Better quality GMRS radios (Allowing external antennas such as the Baofeng GMRS-V1), HAM Radios, and Citizen Band Radios will have better range and communication and therefore better chances to communicate.

The 3-3-3 rules state the following:

1. Turn on your radio every 3 hours (Noon, 3pm, 6pm, 9pm, midnight, 3am, 6am, 9am)
2. Monitor or call for at least 3 minutes (Up to 15 minutes if your battery allows)
3. Tune in on channel 3 (FRS, GMRS, MURS or CB Radio) for your radio type. For HAM radio operators, you should tune to the HAM Calling frequency of 146.52 MHz for 2-Meter band or 446.0 MHz for 70-cm band. After making contact you would then move to another free frequency to continue a conversation.

Wilderness Emergency Protocol

Another emergency protocol which is used mostly in remote wilderness areas where radio cell coverage is sparse or nonexistent is known as the "Wilderness Emergency Protocol". This protocol works like the 3-3-3 protocol in that times and frequencies are defined for use. Under the Wilderness Protocol Ham radio operators outside of repeater range and in questionable cellular telephone coverage are encouraged to monitor standard simplex calling channel VHF (2 Meter) 146.52 MHz, and alternately frequencies 52.525, 223.5, 446.0, and 1294.5 MHz as secondary frequencies depending on the radio bands you have available on your radio. Most handheld ham radios will work 2-meter/70-cm.

This protocol was designed to facilitate HAM radio operators in uninhabited areas outside of repeater range and is strictly focused on HAM Radio channels, not considering FRS, GMRS. MURS, CB Radio or any other type of radio network.

The protocol recommendation is that any station able to do so should monitor the primary frequency (146.52 MHz) and alternate frequencies if possible every three hours starting at 7am for 5 minutes (07:00-07:05 am, then again at 10:00am-10:05am, etc.). If your power on your radio allows it is asked that you monitor every hour instead of every three hours, and if possible even monitor continuously. The protocol calls for monitoring from 7am until 7pm.

Myself as a casual hiker and a ham radio enthusiast, I carry a small portable radio tuned into 146.52 and a second channel such as a repeater that works where I am. If I am hiking with others using GMRS or FRS radios, I may carry a second GMRS radio as well.

12 - HAM Radio – Getting Licensed

For anyone interested in including Ham radio as a part of your emergency preparation plan, first I applaud you. Ham radio will give you more power, range and flexibility over use of GMRS/FRS, MURS, or CB Radio. The use of repeater networks with higher frequency 2-Meter/70-CM Radios, or through use of lower frequency/greater range HF Radio will be a powerful solution for communicating if traditional phones are down.

Ham Licensing is a low-cost solution and starter radios can be purchased for under $100 that will get you on the air. In 2018 the cost of testing is only $15 and goes up minimally each year. Ham Radio is a hobby supported at low cost and high effort by an amazing community. These members of the ham radio community participate in emergency planning at all levels and participate during disasters such as hurricanes providing needed communication in areas otherwise cut off. Ham operators are vital in Puerto Rico as I am writing this after the devastation to the infrastructure following hurricane Maria in September of 2017.

Getting licensed starts out by passing your Technician level license test. Locations and schedules for testing can be found on the ARRL Website at http://www.arrl.org. Passing your test requires the passing of at least 26 questions of 35 total questions on your test which are pulled from a pool of 350 questions. The exact question pool is available from a variety of sources as well as free practice tests and inexpensive books and study sites which can help you pass relatively easily.

The test does require learning some math, but also requires learning of concepts and memorization. Your best method of learning is dependent upon your style of learning, so we won't cover all of that here. For myself as having a degree in electronic engineering much of the math and many of the concepts I already knew. But a great amount of the material can be learned through memorization and study of the concepts and materials.

Is Technician level enough? What else is there?

Great questions! First – Is a technician level license enough? Well, maybe. This level of license will get you legally talking on ham channels and allow you to plug into repeater networks which often cover vast geographic areas. I have participated in conversations and monitored channels used by local EOC centers during hurricanes and weather events. I can talk to others more easily and stay "On the air" while hiking in dead cell zones. A Tech license opens you up to communicating on frequencies with repeaters and short distance radio-to-radio communicating. But what if your planning is to be ready in the event of something happening and in a situation where repeater networks cannot be relied upon?

During Hurricane Irma in September of 2017 in Florida, I could monitor communication that was directly plugged into the EOC of west central Florida. Alerts and bulletins would go out and reports made by hams on the network about conditions witnessed in person where telephones were down. One report I overheard at midnight was of a tornado touch down in Central Florida and occurring as the ham was witnessing it.

But – That situation relied upon a wide repeater network that covers west central Florida. During the hurricane one of the repeater locations experienced power loss and the repeater was running on battery backup. There was no operating generator at that location so it was known that at some point the repeater would power off and those depending on it would be off the air. Other repeater locations (Seven in all) had generators which would run for days or were not experiencing power loss from the local utility.

What about a larger scale event? A worse-case scenario where there was total power loss? Fuel problems, possibly no backup generators.

The next level of Ham radio covers use of lower frequencies starting in the 6 and 10 meter bands and moving up in bands from there. (Remember, higher number in bands means lower frequencies). These lower frequencies experience radio propagation differently and range increases. With a radio running on 10 meters you can easily reach much greater distances than that of 70-cm and 2-meter frequencies.

These upper bands open for legal use to Ham Radio operators with a General or Extra level license. Cost for testing is still low – about $15 (As of 2017/2018). More material to learn, more testing. But more capabilities.

What level you decide ultimately decide to get if you do decide to get licensed should be dependent upon your specific needs. In my own opinion, anyone interested in radio as part of an emergency preparation plan should go for their Technician level license which allows you to legally participate in local nets and use of the radio. Learning through participation and understanding of how radio operates will make the use of radio natural and easier when the skill is needed to put into action.

Appendixes

In the previous sections of our book we touched on a lot of general topics for the newcomer to radio to try to fill in some of the gaps. We hope that we could answer some questions, while we may have at the same time opened some new questions.

The original goal of this book has been to act as a reference guide in print for frequency questions and provide a reference for newer and existing radio enthusiasts, helping to give you a map guide and point you in some of the right directions for better being able to use your radios. In the next appendix pages, you will find specific frequency information for various agencies, radio bands, groups and types of radio use. The table shown below on this page is a description of many of the columns you will find in these tables. Not all columns will be on all the tables. Also, be aware that with many of these frequencies, and many groups and organizations are moving from analog radio into digital and encrypted radio, not all radio frequencies may work in many areas.

Frequency	The repeater output frequency or simplex frequency
License	The FCC Call sign for this frequency
Type	The class station codes for this frequency. B=Base; M=Mobiles/Handhelds; R=Repeater; F=Fixed
Tone	The Sub-audible tone or access code used on the output frequency. PL = CTCSS Tone; DPL = DCS (Digital PL); CSQ = Carrier Squelch/No Tone; NAC = Project 25 Digital Network Access Code; CC = DMR / TRBO Color Code; RAN = NXDN Radio Access Number; DSQ = Yaesu System Fusion Squelch
Alpha Tag	A twelve-character description that is intended as a default alpha tag for scanner listeners.
Description	The description and use of the Frequency
Mode	The Transmission mode used by this frequency
Tag	Service tag for the frequency
Station Class Codes (CS Code)	Describe the purpose for which a station is used.
License	If a license is or is not required
Power	Some radios have specific power restrictions.
Note	Additional note information that may be available for the listing
Band	Band information

Frequency Appendixes: Radio frequency band descriptions and what you would use to access them

In this section of the book we will be providing a list of several differing radio services. These radio services are each accessed through specific types of radio systems, but not any single radio system will be able to access all of these radio frequencies. Below is a quick list of the appendixes which contain frequency services and the types of radios you would generally use to access them.

(Note: All frequency ranges are rounded off here and do not reference specific frequencies. See service details later in this appendix for the given service for more specifics about frequencies used)

Appendix	Radio Service / Description
F – Page 68	**Aircraft / Aviation** Airband – VHF 108-137 MHz, fixed, Aircraft use only, AM ***Radios: AB, FS, SH***
G – Page 70	**American Red Cross - National Frequencies** VHF 121MHz – 135MHz, fixed channels, Restricted use, FM VHF Frequencies used for national emergencies. Many regional and local Red Cross chapters will use UHF or other frequencies unique to their area. Also they leverage use of HAM Radio channels during emergencies. ***Radios: SP, FS, SH***
H – Page 70	**Bureau of Alcohol, Tobacco and Firearms (ATF)** VHF 162MHz – 173MHz fixed channels, Restricted, P25 Can be received on some HAM Radios, Frequency scanners. Digital radios may be needed, signals may be scrambled. ***Radios: FS, SH***
I – Page 71	**Citizens Band (CB) Radio** HF - 26MHz – 27MHz, fixed channels, No license, AM CB Radio frequencies may be received via CB Radios and some frequency scanners. HF Ham Radios may also be able to receive. ***Radios: CB, FS, SH***
J – Page 73	**Civil Air Patrol** VHF 139MHz – 150MHz, Restricted, NFM & P25 CAP uses VHF Radios capable of Narrow FM mode and/or P25 Digital modes. ***Radios: SP, FS, SH***

K – Page 75	**Business Frequencies** VHF 30MHz – UHF 939MHz Various, FM & NFM Frequencies and Channels reserved for business use cover a wide range of frequencies from HF to UHF. Not all radios will be designed to cover all frequencies. ***Radios: SP, FS, SH*, BF****
L – Page 78	**FEMA (Federal Emergency Management Agency)** UHF 406MHz –410MHz, Restricted, P25 ***Radios: FS, SP, SH****
M – Page 79	**FRS (Family Radio Service)** UHF 462MHz – 467MHz, Fixed channels, NFM All frequencies shared with GMRS, Max 2 watts on SOME channels, ½ watt on other channels ***Radios: SH, FS, BF*, FRS***
N – Page 81	**GMRS (General Mobile Radio Service)** UHF 462MHz – 467MHz, GMRS License, Fixed Freq, NFM Some frequencies shared with FRS, Max 50 watts on SOME channels, most channels max at 5 watts ***Radios: GMRS, FS, SH*, BF****
O – Page 86	**Ham Radio Simplex Radio Frequencies** HF – UHF Frequencies, Ham License, Multiple Modes The full spectrum of frequencies available to Amateur Radio licenses, various powers based on frequencies and license level ***Radios: Various Ham Radios, FS, BF***
P – Page 87	**International Space Station** VHF 145MHz, UHF 437MHz, Ham License, FM & Telem Four specific frequencies available to Ham operators to speak to astronauts about the ISS ***Radios: Various Ham Radios, FS, BF***
Q – Page 88	**Itinerant Business Radio** VHF and UHF, Various Frequencies, AM & NFM Modes ***Radios: SP, SH, BF****
R – Page 89	**Marine Radio** VHF 156MHz – 162MHz, Marine Use only, FM License NOT required but restricted to Marine use only ***Radios: SP Marine VHF, FS, BF****
S – Page 93	**Military Radio** HF and UHF Mult Frequencies, AM/FM Modes, Restricted use ***Radios: FS, SP, SH***
T – Page 96	**MURS Public Use Radio** VHF 151MHz – 154Mh, No license, FM & NFM, No license ***Radios: MURs, FS. SH*, BF****
U – Page 97	**National Disaster Medical** UHF 406MHz – 412MHz, restricted, P25 ***Radios: SP, SH*. FS***
V – Page 98	**Public Safety and Law Enforcement** UHF 406MHz – Telemetry ***Radios: SP, FS***

W – Page 99	**UHF Federal Law Enforcement Interoperability** UHF 410MHz – 414MHz, Restricted, NFM and P25 ***Radios: SP, FS***
X – Page 100	**UHF Medical** UHF Frequencies 462MHz – 463MHz, NFM, Restricted use ***Radios: SP,***
Y - Page 102	**UHF National Interoperability** UHF Frequencies 453MHz, NFM, Restricted use ***Radios: SP, SH*, FS, BF****
Z – Page 102	**VHF Federal Incident Response** VHF 169MHz – 173MHz, NFM, Restricted use ***Radios: SP, SH*, BF, FS***
AA – Page 103	**USAR (Urban Search and Rescue)** UHF 408MHz – 412MHz, NFM, P25, Restricted ***Radios: SP, SH*, FS***
AB – Page 104	**VHF Law Enforcement** VHF Frequencies 167MHz – 168MHz, NFM, P25, Restricted ***Radios: SP, SH*, BF*, FS***
AC – Page 104	**VHF Low-Band Public Services** HF 39MHz & 45MHz, Fire & Law Enforcement, FM, Restricted use only ***Radios: SP, SH, FS***
AD – Page 105	**VHF National Interoperability** VHF 151MHz – 161MHz, NFM, Restricted ***Radios: SP, SH*, BF*, FS***
AE – Page 106	**VHF Public Safety** VHF 154MHz – 155MHz, NFM, Restricted Frequencies here used for Fire, Search/Rescue, EMS, Police ***Radios: FP. FS, SH*, BF****
AF – Page 107	**Weather Radio (Broadcast only – Not used for 2-way radio)** VHF 162MHz, 7 Channels, FM, Receive only ***Radios: FS, Weather Radios, SH*, BF****
AI – Page 108	**VHF & UHF Television Channel Audio Frequencies** Listening to Television Audio is available on some types of radios and/or scanners. This is a list of those audio frequencies.
AI – Page 111	**Shortwave Radio** Shortwave radio uses frequencies generally from 1.6MHz to 30MHz and can be received by Shortwave radio receivers as well as some Ham radios. ***Radios: Shortwave, SH****

See page 57 for Key to codes in this table

Key to Abbreviations in this table on pages 54-56:

AB	Air Band (Aircraft radio)
FS	Frequency Scanners which can receive but not transmit
SH	Some Ham Radios, depending upon bands the radios support
SH*	Some Ham Radios can receive and transmit on these frequencies which are outside of legal use for amateur radio.
SP	Radios specifically programmed for support of these frequencies/channels.
CB	Citizens Band Radio – No license required in the US
BF	Baofeng or similar 2-meter/70-cm radios.
BF*	Baofeng and similar radios designed for ham use can receive these frequencies but cannot be used legally for transmitting as they may create interference and may not be following FCC Restrictions for these frequencies. Baofeng and similar radios may work in 136MHz-174MHz and 400MHz-520MHz which are outside of US FCC Bands established for HAM Radio use.
FM	Freuency Modulated Mode
NFM	Narrow Frequency Modulated Mode – Uses a smaller bandwidth for modulation
AM	Amplitude Modulation (CB Radio for instance)
P25	Mode of digital radio used for public service.

A: National Repeater Councils

If you use a HAM Radio for two-way communicating (License required) or just use it to listen in on for conversation and news information, then having a printed reference to the local repeaters in your area and the band plan for your region or state is critical in the event of an emergency that results in power and/or communications and cell system loss. The national frequencies we have provided in this book are great reference sources for general emergency and government frequencies that you can monitor during such an event, but each region or state has lists of repeaters and radio frequencies that have been designated for types of use by organizations that are far too in depth and change too frequently for us to update in this book.

By visiting the Frequency Coordinator organizations for your own state or region you should be able to obtain the plans that are specific to your area. We recommend you go to the appropriate website to find and print those lists and keep them in a folder with your emergency supplies. The list updates frequently enough that an annual refresh of your list should be sufficient. If you are or become an active Ham radio participant, you will become familiar with some of the repeater networks and emergency communication networks in your area. Otherwise having the printed frequencies and contacts for your area will be even more important. For some states or regions no websites were available however specific named contacts with email addresses are sometimes available from the National Frequency Coordinators Council's (NFCC) own website.

National Frequency Coordinators Council - http://www.nfcc.us/

State	Council Name
Alabama	Alabama Repeater Council, Inc http://www.alabamarepeatercouncil.org/
Alaska	Alaska Repeater Coordinators No Web Address
Arizona	Amateur Radio Council of Arizona http://www.azfreqcoord.org/
Arkansas	Arkansas Repeater Council http://www.arkansasrepeatercouncil.org/
California	The Northern Amateur Relay Council of California (NARCC) http://www.narcc.org/
California	Southern California Repeater & Remote Base Association http://www.scrrba.org/
California	Two Meter Area Spectrum Management Association http://www.tasma.org/
California	220 MHz Spectrum Management Association of Southern California http://www.220sma.org/
Colorado	Colorado Council of Amateur Radio Clubs http://www.ccarc.net/
Connecticut	Connecticut Spectrum Management Association, Inc. http://www.ctspectrum.com/
Delaware	The Mid-Atlantic Repeater Council, Inc. (T-MARC) http://www.t-marc.org/
District of Columbia	The Mid-Atlantic Repeater Council, Inc. (T-MARC) http://www.t-marc.org/
Florida	Florida Repeater Council http://florida-repeaters.org/
Georgia	South Eastern Repeater Association, Inc. (SERA) http://sera.org/index.php/home/sera-districts/sera-districts-ga/
Hawaii	HAWAII - Hawaii State Repeater Advisory Council (HSRAC) No Web Address
Idaho	SEICC No Web Address
Illinois	Illinois Repeater Association, Inc. (IRA) http://www.ilra.net/
Indiana	Indiana Repeater Council http://www.ircinc.org/
Iowa	Iowa Repeater Council http://www.iowarepeater.org/

State	Council Name
Kansas	South Eastern Repeater Association, Inc. (SERA) http://www.ksrepeater.com/
Kentucky	South Eastern Repeater Association, Inc. (SERA) http://sera.org/index.php/home/sera-districts/sera-districts-ky/
Louisiana	LCARC- Louisiana Council of Amateur Radio Clubs http://www.lacouncil.net/
Maine	New England Spectrum Management Council (NESMC) http://www.nesmc.org/
Maryland	The Mid-Atlantic Repeater Council, Inc. (T-MARC) http://www.tmarc.org/
Massachusetts	New England Spectrum Management Council (NESMC) http://www.nesmc.org/
Michigan	an Area Repeater Council (MARC) 'www.qsl.net/miarc/
Minnesota	Minnesota Repeater Council (MRC) http://www.mrc.gen.mn.us/
Mississippi	South Eastern Repeater Association, Inc. (SERA) http://sera.org/index.php/home/sera-districts/sera-districts-mi/
Missouri	Missouri Repeater Council, Inc. (MRC) http://www.missourirepeater.org/
Nevada	Northern Nevada - CARCON http://www.carcon.org/
Nevada	Southern Nevada - Southern Nevada Repeater Council (SNRC)
New Hampshire	New England Spectrum Management Council (NESMC) http://www.nesmc.org/
New Jersey	Metropolitan Coordination Association, Inc. (MetroCor) http://www.metrocor.net/
New Jersey	Southern/Western New Jersey http://www.arcc-inc.org/
New Mexico	New Mexico Frequency Coordination Committee (NMFCC) No Web Address
New York	NY City & Long Island - Metro Coordination Association, Inc. http://www.metrocor.net/
New York	NE New York - VT Independent Repeater Coordinating Committee http://www.ranv.org/rptr.html
New York	North Central New York - Saint Lawrence Valley Repeater Council http://www.slvrc.org/
New York	Upstate (Mid) New York http://www.unyrepco.org/

State	Council Name
New York	Western New York http://www.wnysorc.org/
North Carolina	South Eastern Repeater Association, Inc. (SERA) http://sera.org/index.php/home/sera-districts/sera-districts-nc/
South Carolina	South Eastern Repeater Association, Inc. (SERA) http://sera.org/index.php/home/sera-districts/sera-districts-sc/
Tennessee	South Eastern Repeater Association, Inc. (SERA) http://sera.org/index.php/home/sera-districts/sera-districts-tn/
Virginia	South Eastern Repeater Association, Inc. (SERA) http://sera.org/index.php/home/sera-districts/sera-districts-va/
West Virginia	South Eastern Repeater Association, Inc. (SERA) http://sera.org/index.php/sera-districts-wv/
Ohio	Ohio Area Repeater Council (OARC) http://www.oarc.com/
Oklahoma	Oklahoma Repeater Society Inc. (ORSI) http://www.oklahomarepeatersociety.org/
Oregon	Oregon Region Relay Council, Inc. (ORRC) http://www.orrc.org/
Pennsylvania	Area Repeater Coordination Council, Inc. (ARCC) http://www.arcc-inc.org/
Pennsylvania	Western - Western Pennsylvania Repeater Council (WPRC) http://www.qsl.net/wprc
Rhode Island	New England Spectrum Management Council (NESMC) http://www.nesmc.org/
Texas	Texas VHF-FM Society (TVHFS) http://www.txvhffm.org/
Utah	UVHFS http://www.ussc.com/~uvhfs/
Vermont	VT Independent Repeater Coordinating Committee (VIRCC) No Web Address
Washington	W. Washington Amateur Relay Association, Inc. (WWARA) http://www.wwara.org/
Wisconsin	Wisconsin Association of Repeaters (WAR) http://www.wi-repeaters.org/
Wyoming	Wyoming Council of Radio Clubs (WCARC) http://www.n7gt.com/coordination.html

B: Sample Frequency Band Plan (Florida)

The following is the Band Plan assigned by the Florida Repeater Council.

Band	Frequency	Usage
2-Meter	144.000 - 144.050	CW/EME
2-Meter	144.050 - 144.100	CW and Weak Signal
2-Meter	144.100 - 144.200	CW and Weak Signal SSB
2-Meter	144.200	SSB Calling Frequency
2-Meter	144.200 - 144.275	SSB
2-Meter	144.275 - 144.300	Propagation Beacons
2-Meter	144.300 - 144.500	Satellite/OSCAR
2-Meter	144.500 - 144.900	FM Repeater Inputs
2-Meter	144.900 - 145.100	Weak Signal, FM Simplex, Digital/Packet
2-Meter	145.100 - 145.500	FM Repeater Outputs
2-Meter	145.500 - 145.800	Miscellaneous and Experimental Modes
2-Meter	145.800 - 146.000	Satellite/OSCAR
2-Meter	146.010 - 146.370	FM Repeater Inputs
2-Meter	146.400 - 146.580	FM Simplex
2-Meter	146.610 - 147.390	FM Repeater Outputs
2-Meter	147.420 - 147.570	FM Simplex
2-Meter	147.600 - 147.990	FM Repeater Inputs
1.25-M	222.000 - 222.025	EME
1.25-M	222.050 - 222.060	Propagation Beacons
1.25-M	222.1	CW and SSB Calling Frequency
1.25-M	222.100 - 222.150	CW and SSB Weak Signal
1.25-M	222.150 - 222.250	Local Coordinators Option
1.25-M	222.250 - 223.380	FM Repeater Inputs
1.25-M	223.400 - 223.520	FM Voice Simplex
1.25-M	223.520 - 223.640	Digital/Packet
1.25-M	223.640 - 223.700	Links and Control
1.25-M	223.710 - 223.850	Local Coordinators Option
1.25-M	223.850 - 224.980	FM Repeater Outputs
70-cm	420.000 - 426.000	ATV, Control Links and Experimental
70-cm	426.000 - 432.000	ATV Simplex
70-cm	430.050 - 430.950	Digital/Packet, 100 khz bandwidth channels
70-cm	431.025 - 431.950	Digital/Packet, 25 khz bandwidth channel
70-cm	432.000 - 432.070	EME
70-cm	432.070 - 432.100	CW Weak Signal
70-cm	432.100	SSB Calling Frequency
70-cm	432.100 - 432.300	Mixed Mode and Weak Signal

Band	Frequency	Usage
70-cm	432.300 - 432.400	Propagation Beacons
70-cm	432.400 - 433.000	Mixed Mode and Weak Signal
70-cm	433.000 - 435.000	Auxiliary and Repeater Links
70-cm	435.000 - 438.000	Satellite
70-cm	438.000 - 444.000	ATV Repeater Input and FM Repeater Links
70-cm	440.975 - 441.075	Digital/Packet, 25 khz bandwidth channel
70-cm	442.000 - 445.000	FM Repeater Inputs and Outputs (local option)
70-cm	445.000 - 447.000	Auxiliary and Control Links, Simplex, FM Repeater Inputs and Outputs (local option)
70-cm	447.000 - 450.000	FM Repeater Inputs and Outputs (local option)

C: Phonetic Alphabet

Following is the NATO Standard Phonetic alphabet used when communicating call signs across the radio. Use of this alphabet makes it easier to understand characters where individually spoken multiple characters may sound very similar.

Letter	NATO		Letter	NATO
A	Alpha		N	November
B	Bravo		O	Oscar
C	Charlie		P	Papa
D	Delta		Q	Quebec
E	Echo		R	Romeo
F	Foxtrot		S	Sierra
G	Golf		T	Tango
H	Hotel		U	Uniform
I	India		V	Victor
J	Juliet		W	Whiskey
K	Kilo		X	X-ray
L	Lima		Y	Yankee
M	Mike		Z	Zulu

D: Ham Radio Frequencies and Calling Frequencies

The following chart are the frequency ranges of the most common Ham Radio bands. Not all levels of Ham licenses are permitted us of all frequencies and many of the frequencies have restrictions between use of FM, SSB (Single Side Band) and CW (Morse Code) modes.

BAND	Freq Low	Freq High		Notes
23 cm	1240 MHz	1300 MHz	UHF	Local & Line of sight
33 cm	902.0 MHz	928.0 MHz	UHF	Local & Line of sight
70 cm	420.0 MHz	450.0 MHz	UHF	Local & Line of sight
1.25 M	219.0 MHz	220.0 MHz	VHF	Local & Line of sight
2 M	144.0 MHz	450 MHz	VHF	Local & Line of sight
6 M	50.0 MHz	54.0 MHz	VHF	Local & Line of sight
10 M	28.0 MHz	29.7 MHz	HF	Long Distance
12 M	24.89 MHz	24.99 MHz	HF	Long Distance
15 M	21.0 MHz	21.45 MHz	HF	Long Distance
17 M	18.068 MHz	18.168 MHz	HF	Long Distance
20 M	14.0 MHz	14.35 MHz	HF	Long Distance
30 M	10.1 MHz	10.15 MHz	HF	Long Distance
40 M	7.0 MHz	7.3 MHz	HF	Long Distance
60 M	5330.5 KHz	5403.5 KHz	HF	Long Distance
80 M	3.5 MHz	4.0 MHz	HF	Long Distance
160 M	1.8 MHz	2.0 MHz	MF	Long Distance

Ham radio licenses are broken into three levels:

- **Technician**: The entry level of license which allows voice (FM) and CW use of some of the frequencies. A majority of radio available to Technician class holders occurs in the 70cm and 2-Meter bands which is line of sight.
- **General License**: This second-tier license level opens up most of the bands to the license holder allowing for long-range communication over the higher bands.
- **Amateur Extra**: Addition of a small number of exclusive frequency ranges on the 15, 20, 40 and 80 meter bands.

The following frequencies are established common and calling frequencies used on various bands for Ham radio.

Band	Frequency	Description
160 Meters	1.800 - 2.000 MHz	CW (Morse Code)
160 Meters	1.81 MHz	CW QRP
160 Meters	1.91 MHz	SSB QRP
80 Meters	3.790-3.800 MHz	DX window
80 Meters	3.845 MHz	SSTV
80 Meters	3.885 MHz	AM calling frequency
60 Meters	5403.5 MHz	USB phone and CW/RTTY/data
40 Meters	7.29 MHz	AM calling frequency
20 Meters	14.286 MHz	AM calling frequency
10 Meters	28.150-28.190 MHz	CW (Morse Code)
10 Meters	28.300-29.300 MHz	Phone
10 Meters	29.000-29.200 MHz	AM
10 Meters	29.6 MHz	FM Simplex
6 Meters	50.1-50.3 MHz	SSB, CW (Morse Code)
6 Meters	50.10-50.125 MHz	DX window
6 Meters	50.125 MHz	SSB calling
6 Meters	50.62 MHz	Digital (packet) calling
6 Meters	52.525 MHz	Primary FM simplex
6 Meters	52.54 MHz	Secondary FM simplex
6 Meters	53.02 MHz	Simplex
6 Meters	53.52, 53.9 MHz	Simplex
2 Meters	144.2 MHz	National calling frequency
2 Meters	146.52 MHz	National Simplex Calling Frequency
2 Meters	147.42-147.57 MHz	Simplex
1.25 Meters	222.1 MHz	SSB & CW calling frequency
1.25 Meters	223.40-223.52 MHz	FM simplex
70 Centimeters	432.1 MHz	70-cm calling frequency
70 Centimeters	446 MHz	National simplex frequency

E: Frequency Recording Logs

Use the following forms to manually record details about frequency channels you wish to record for use on your Ham radio.

Channel Name:	
Frequency:	
Tone Mode:	
Tone SQL:	
Cross Mode:	
DTCS Code:	
RX DTCS Code:	
DTCS Pol:	
Duplex:	
Offset:	
Mode:	

Channel Name:	
Frequency:	
Tone Mode:	
Tone SQL:	
Cross Mode:	
DTCS Code:	
RX DTCS Code:	
DTCS Pol:	
Duplex:	
Offset:	
Mode:	

Channel Name:	
Frequency:	
Tone Mode:	
Tone SQL:	
Cross Mode:	
DTCS Code:	
RX DTCS Code:	
DTCS Pol:	
Duplex:	
Offset:	
Mode:	

Channel Name:	
Frequency:	
Tone Mode:	
Tone SQL:	
Cross Mode:	
DTCS Code:	
RX DTCS Code:	
DTCS Pol:	
Duplex:	
Offset:	
Mode:	

Note: *Not all fields may be needed by a radio channel.*

F: AIRCRAFT / AVIATION

Frequency	Description	Alpha Tag	Band	Mode	Tag	Tone	Type
121.5000	Aircraft Emergency / Distress (VHF Guard)	VHF Guard	Airband	AM	Aircraft		M
121.7750	Emergency Locator Transmitter (ELT) Training Beacons	ELT Training	Airband	AM	Aircraft		M
121.9500	Aviation Support	AvSup 121.95	Airband	AM	Business		M
122.0000	Flight Watch Weather	Flt Watch WX	Airband	AM	Aircraft		M
122.7000	Unicom	FAA 122.700	Airband	AM	Aircraft		M
122.7250	Unicom	FAA 122.725	Airband	AM	Aircraft		M
122.7500	Fixed wing aircraft air-to-air	Fixed Air-Air	Airband	AM	Aircraft		M
122.7750	Aviation Support	AvSup 122.775	Airband	AM	Business		M
122.8000	Unicom	FAA 122.8	Airband	AM	Aircraft		M
122.8500	Multicom, Aviation Support	Multi 122.85	Airband	AM	Aircraft		M
122.9000	Multicom	FAA 122.900	Airband	AM	Aircraft		M
122.9000	Multicom	FAA 122.900	Airband	AM	Aircraft		M
122.9250	Multicom	FAA 122.925	Airband	AM	Aircraft		M
122.9500	Unicom - Controlled Airports	Unicom 122.95	Airband	AM	Aircraft	CSQ	BM
122.9750	Unicom	FAA 122.975	Airband	AM	Aircraft		M
123.0000	Unicom	FAA 123.000	Airband	AM	Aircraft		M
123.0250	Helicopter Air-to-Air	Heli Air-Air	Airband	AM	Aircraft		M
123.0500	Unicom	FAA 123.05	Airband	AM	Aircraft		M
123.0750	Unicom	FAA 123.075	Airband	AM	Aircraft		M
123.1000	Search and Rescue primary, ATC for special events secondary	SAR Primary	Airband	AM	Aircraft		M
123.1250	Flight Test itinerant	FlightTest123.12	Airband	AM	Business		M

Frequency	Description	Alpha Tag	Band	Mode	Tag	Tone	Type
123.1500	Flight Test itinerant	FlightTest123.15	Airband	AM	Business		M
123.1750	Flight Test itinerant	FlightTest123.17	Airband	AM	Business		M
123.2000	Flight Test	FlightTest123.2	Airband	AM	Business		M
123.2250	Flight Test	FlightTest123.22	Airband	AM	Business		M
123.2500	Flight Test	FlightTest123.25	Airband	AM	Business		M
123.2750	Flight Test	FlightTest123.27	Airband	AM	Business		M
123.3000	Aviation Support	AvSup 123.3	Airband	AM	Business		M
123.3250	Flight Test	FlightTest123.32	Airband	AM	Business		M
123.3500	Flight Test	FlightTest123.35	Airband	AM	Business		M
123.3500	Flight Test	FlightTest123.35	Airband	AM	Business		M
123.3750	Flight Test	FlightTest123.37	Airband	AM	Business		M
123.4000	Flight Test itinerant	FlightTest123.4	Airband	AM	Business		M
123.4250	Flight Test itinerant	FlightTest123.42	Airband	AM	Business		M
123.4500	Flight Test/Unofficial Air-to-Air	FlightTest123.45	Airband	AM	Business		M
123.4750	Flight Test	FlightTest123.47	Airband	AM	Business		M
123.5000	Aviation Support	AvSup 123.5	Airband	AM	Business		M
123.5250	Flight Test	FlightTest123.52	Airband	AM	Business		M
123.5500	Flight Test	FlightTest123.55	Airband	AM	Business		M
123.5750	Flight Test	FlightTest123.57	Airband	AM	Business		M
126.2000	Military Common (advisory)	MilCom 126.2	Airband	AM	Aircraft		M
129.5250	Deicing Common	Deicing Comm	Airband	AM	Business		M
134.1000	Military Common (advisory)	MilCom 134.1	Airband	AM	Aircraft		M
135.8500	FAA Flight Inspection	FltIns135.85	Airband	AM	Federal		M

G: AMERICAN RED CROSS – National Frequencies

The following are NATIONAL frequencies allocated for use by the American Red Cross. Other frequencies are also allocated at state or regional levels.

Frequency	Description	Alpha Tag	Comment	Mode	Tag	Type
47.4200	Primary Nationwide	REDCROSS 1	Primary Nationwide	FM	Emergency Ops	BM
47.5000	Secondary Nationwide	REDCROSS 2	Secondary Nationwide	FM	Emergency Ops	BM

H: Bureau of Alcohol, Tobacco and Firearms (ATF)

Frequency	Description	Alpha Tag	Band	Mode	Tag	Tone	Type
165.2875	Nets (various input NACs)	ATF Nets	Government	P25	Federal	650 NAC	RM
165.9125	Tactical 5	ATF Tac 5	Government	P25	Federal	650 NAC	M
167.1625	Tactical 1	ATF Tac 1	Government	P25	Federal	650 NAC	M
168.3500	Tactical 4	ATF Tac 4	Government	P25	Federal	650 NAC	M
170.9625	Tactical 2	ATF Tac 2	Government	P25	Federal	650 NAC	M
173.0125	Tactical 3	ATF Tac 3	Government	P25	Federal	650 NAC	M
173.8875	Tactical 6	ATF Tac 6	Government	P25	Federal	650 NAC	M

I: Citizens Band (CB) Radio

Citizens Band Radio consists of 40 designated channels in the 26/27 MHz frequency ranges with a max power level of 4 watts for standard use, or 12 watts if Single-Side band (SSB) is used. CB Radio does not require a license to operate. The radio signal wavelength is 11 meters (36 Feet). A quarter wave antenna will run a length of 2.7 Meters, or roughly 9 feet. Due to the longer antenna length CB Radio is not as popular or common for handheld radio communication.

Frequency	Description	Band	Mode	Tone	Type	License	Power
26.9650	Channel 01	Citizens Band	AM	CSQ	M	No	4 watts (12 SSB)
26.9750	Channel 02	Citizens Band	AM	CSQ	M	None Needed	4 watts (12 SSB)
26.9850	Channel 03	Citizens Band	AM	CSQ	M	None Needed	4 watts (12 SSB)
27.0050	Channel 04	Citizens Band	AM	CSQ	M	None Needed	4 watts (12 SSB)
27.0150	Channel 05	Citizens Band	AM	CSQ	M	None Needed	4 watts (12 SSB)
27.0250	Channel 06	Citizens Band	AM	CSQ	M	None Needed	4 watts (12 SSB)
27.0350	Channel 07	Citizens Band	AM	CSQ	M	None Needed	4 watts (12 SSB)
27.0550	Channel 08	Citizens Band	AM	CSQ	M	None Needed	4 watts (12 SSB)
27.0650	Channel 09 - Emergency	Citizens Band	AM	CSQ	M	None Needed	4 watts (12 SSB)
27.0750	Channel 10	Citizens Band	AM	CSQ	M	None Needed	4 watts (12 SSB)
27.0850	Channel 11	Citizens Band	AM	CSQ	M	None Needed	4 watts (12 SSB)
27.1050	Channel 12	Citizens Band	AM	CSQ	M	None Needed	4 watts (12 SSB)
27.1150	Channel 13	Citizens Band	AM	CSQ	M	None Needed	4 watts (12 SSB)
27.1250	Channel 14	Citizens Band	AM	CSQ	M	None Needed	4 watts (12 SSB)

27.1350	Channel 15	Citizens Band	AM	CSQ	M	None Needed	4 watts (12 SSB)
27.1550	Channel 16	Citizens Band	AM	CSQ	M	None Needed	4 watts (12 SSB)
27.1650	Channel 17	Citizens Band	AM	CSQ	M	None Needed	4 watts (12 SSB)
27.1750	Channel 18	Citizens Band	AM	CSQ	M	None Needed	4 watts (12 SSB)
27.1850	Channel 19 - Highway Traffic	Citizens Band	AM	CSQ	M	None Needed	4 watts (12 SSB)
27.2050	Channel 20	Citizens Band	AM	CSQ	M	None Needed	4 watts (12 SSB)
27.2150	Channel 21	Citizens Band	AM	CSQ	M	None Needed	4 watts (12 SSB)
27.2250	Channel 22	Citizens Band	AM	CSQ	M	None Needed	4 watts (12 SSB)
27.2350	Channel 24	Citizens Band	AM	CSQ	M	None Needed	4 watts (12 SSB)
27.2450	Channel 25	Citizens Band	AM	CSQ	M	None Needed	4 watts (12 SSB)
27.2550	Channel 23	Citizens Band	AM	CSQ	M	None Needed	4 watts (12 SSB)
27.2650	Channel 26	Citizens Band	AM	CSQ	M	None Needed	4 watts (12 SSB)
27.2750	Channel 27	Citizens Band	AM	CSQ	M	None Needed	4 watts (12 SSB)
27.2850	Channel 28	Citizens Band	AM	CSQ	M	None Needed	4 watts (12 SSB)
27.2950	Channel 29	Citizens Band	AM	CSQ	M	None Needed	4 watts (12 SSB)
27.3050	Channel 30	Citizens Band	AM	CSQ	M	None Needed	4 watts (12 SSB)
27.3150	Channel 31	Citizens Band	AM	CSQ	M	None Needed	4 watts (12 SSB)
27.3250	Channel 32	Citizens Band	AM	CSQ	M	None Needed	4 watts (12 SSB)
27.3350	Channel 33	Citizens Band	AM	CSQ	M	None Needed	4 watts (12 SSB)
27.3450	Channel 34	Citizens Band	AM	CSQ	M	None Needed	4 watts (12 SSB)
27.3550	Channel 35	Citizens Band	AM	CSQ	M	None Needed	4 watts (12 SSB)
27.3650	Channel 36	Citizens Band	AM	CSQ	M	None Needed	4 watts (12 SSB)

27.3750	Channel 37	Citizens Band	AM	CSQ	M	None Needed	4 watts (12 SSB)
27.3850	Channel 38	Citizens Band	AM	CSQ	M	None Needed	4 watts (12 SSB)
27.3950	Channel 39	Citizens Band	AM	CSQ	M	None Needed	4 watts (12 SSB)
27.4050	Channel 40	Citizens Band	AM	CSQ	M	None Needed	4 watts (12 SSB)

J: Civil Air Patrol

The Civil Air Patrol is a federally supported non-profit organization serving as an official civilian auxiliary of the US Air Force comprised of volunteers. The mission of the Civil Air Patrol is to provide emergency services including search and rescue operations, education for youth and the public, and cadet programs for teenage youths. The CAP also helps with various government and private agencies including law enforcement and the American Red Cross.

Frequency	Description	Alpha Tag	Mode	Tag	Tone	Type
139.8750	Tactical 1 Analog	CAP Tac 1 A	NFM	Military	173.8 PL	BM
139.8750	Tactical 1 Digital	CAP Tac 1 D	P25	Military	6CA NAC	BM
141.0000	Command and Control 2 Analog	CAP CC 2 A	NFM	Military	131.8 PL	BM
141.0000	Command and Control 2 Digital	CAP CC 2 D	P25	Military	526 NAC	BM
141.5750	Command and Control 1 Digital	CAP CC 1 D	P25	Military	4F9 NAC	BM
141.5750	Command and Control 1 Analog	CAP CC 1 A	NFM	Military	127.3 PL	BM
148.1250	Ch R63 Airborne Portable Tactical Repeater Digital	CAP R63 D	P25	Military	7F3 NAC	RM
148.1250	Ch R63 Airborne Portable Tactical Repeater Analog	CAP R63 A	NFM	Military	203.5 PL	RM
148.1250	Ch R01R61 Odd Primary Repeater Digital	CAP PriRptrD	P25	Military		RM
148.1250	Ch R01R61 Odd Primary Repeater Analog	CAP PriRptrA	NFM	Military		RM

Frequency	Description	Alpha Tag	Mode	Tag	Tone	Type
148.1375	Ch R68 Portable and Aero Repeater Digital	CAP R68 D	P25	Military	788 NAC	RM
148.1375	Ch R70 Portable and Aero Repeater Digital	CAP R70 D	P25	Military	656 NAC	RM
148.1375	Ch R70 Portable and Aero Repeater Analog	CAP R70 A	NFM	Military	162.2 PL	RM
148.1375	Ch R69 Portable and Aero Repeater Digital	CAP R69 D	P25	Military	526 NAC	RM
148.1375	Ch R69 Portable and Aero Repeater Analog	CAP R69 A	NFM	Military	131.8 PL	RM
148.1375	Ch R67 Portable and Aero Repeater Analog	CAP R67 A	NFM	Military	203.5 PL	RM
148.1375	Ch R67 Portable and Aero Repeater Digital	CAP R67 D	P25	Military	7F3 NAC	RM
148.1375	Ch R68 Portable and Aero Repeater Analog	CAP R68 A	NFM	Military	192.8 PL	RM
148.1500	Ch R64 Airborne Portable Tactical Repeater Analog	CAP R64 A	NFM	Military	203.5 PL	RM
148.1500	Ch R64 Airborne Portable Tactical Repeater Digital	CAP R64 D	P25	Military	7F3 NAC	RM
148.1500	Ch R02R62 Even Secondary Repeater Digital	CAP	P25	Military		RM
148.1500	Ch R02R62 Even Secondary Repeater Analog	CAP	NFM	Military		RM
149.2750	Air 1 Digital	CAP Air 1 D	P25	Military	585 NAC	BM
149.2750	Air 1 Analog	CAP Air 1 A	NFM	Military	141.3 PL	BM
150.2250	Guard 1 Digital	CAP Guard1	P25	Military	656 NAC	BM
150.2250	Guard 1 Analog	CAP Guard1	NFM	Military	162.2 PL	BM
150.5625	Air 2 Digital	CAP Air 2 D	P25	Military	5EA NAC	BM
150.5625	Air 2 Analog	CAP Air 2 A	NFM	Military	151.4 PL	BM

K: Business Frequencies

Following are frequencies which have been set aside for dedicated business radio use. Radio frequencies being used include those across several bands ranging from the 30 MHz range in the lower VHF band up to radio in the 900 UHF Bands. MURS Radio was carved out of 5 frequencies in the business radio use and in some cases those frequencies may still be used by businesses who hold grandfathered licenses for their use. In cases where a business is using a MURS frequency that the public is trying to use, the priority always goes to the business with the grandfathered use license.

Frequency	Description	Alpha Tag	Mode	Tag	Type
30.8400	Business	Business	FM	Business	M
33.1200	Business	Business	FM	Business	M
33.1400	Business	Business	FM	Business	M
33.4000	Business	Business	FM	Business	M
35.0200	Business	Business	FM	Business	M
35.0400	Business	Business	FM	Business	M
42.9800	Business Pink Dot	Pink Dot	FM	Business	M
43.0400	Business	Business	FM	Business	M
151.5050	Business	Business	NFM	Business	M
151.5125	Business	Business	NFM	Business	M
151.6250	Business Red Dot	Red Dot	NFM	Business	M
151.6400	Business	Business	NFM	Business	M
151.7000	Business	Business	NFM	Business	M
151.7600	Business	Business	NFM	Business	M

Frequency	Description	Alpha Tag	Mode	Tag	Type
151.9550	Business Purple Dot	Purple Dot	NFM	Business	M
154.5150	Business	Business	NFM	Business	M
154.5275	Business	Business	NFM	Business	M
154.5400	Business	Business	NFM	Business	M
158.4000	Business	Business	NFM	Business	M
451.8000	Business	Business	NFM	Business	M
451.8125	Business	Business	NFM	Business	M
456.8000	Business	Business	NFM	Business	M
456.8125	Business	Business	NFM	Business	M
457.5250	Business	Business	NFM	Business	M
457.5375	Business	Business	NFM	Business	M
457.5500	Business	Business	NFM	Business	M
457.5625	Business	Business	NFM	Business	M
457.5750	Business	Business	NFM	Business	M
457.5875	Business	Business	NFM	Business	M
457.6000	Business	Business	NFM	Business	M
457.6125	Business	Business	NFM	Business	M
464.5000	Business Brown Dot	Brown Dot	NFM	Business	M
464.5500	Business Yellow Dot	Yellow Dot	NFM	Business	M
467.7500	Business	Business	NFM	Business	M
467.7625	Business J Dot	J Dot	NFM	Business	M

Frequency	Description	Alpha Tag	Mode	Tag	Type
467.7750	Business	Business	NFM	Business	M
467.7875	Business	Business	NFM	Business	M
467.8000	Business	Business	NFM	Business	M
467.8125	Business K Dot	K Dot	NFM	Business	M
467.8250	Business	Business	NFM	Business	M
467.8375	Business	Business	NFM	Business	M
467.8500	Business Silver Star	Silver Star	NFM	Business	M
467.8625	Business	Business	NFM	Business	M
467.8750	Business Gold Star	Gold Star	NFM	Business	M
467.8875	Business	Business	NFM	Business	M
467.9000	Business Red Star	Red Star	NFM	Business	M
467.9125	Business	Business	NFM	Business	M
467.9250	Business Blue Star	Blue Star	NFM	Business	M
469.5000	Business	Business	NFM	Business	M
469.5500	Business	Business	NFM	Business	M
808.4875	Business	Business	FM	Business	M
853.4875	Business unofficial but widely used itinerant	Business	FM	Business	M
897.6375	Business	Business	NFM	Business	M
897.6625	Business	Business	NFM	Business	M
900.9750	Business	Business	NFM	Business	M
900.9875	Business	Business	NFM	Business	M

Frequency	Description	Alpha Tag	Mode	Tag	Type
936.6375	Business	Business	NFM	Business	M
936.6625	Business	Business	NFM	Business	M
939.9750	Business	Business	NFM	Business	M
939.9875	Business	Business	NFM	Business	M

L: FEMA (Federal Emergency Management Agency)

Frequency	Description	Alpha Tag	Band	Mode	Tag	Tone	Type	License
406.2625	Ch. 1/2	FEMA 1/2	Government	P25	Federal		RM	
407.0625	Ch. 3/4	FEMA 3/4	Government	P25	Federal		RM	
407.4500	Ch. 11/12	FEMA 11/12	Government	P25	Federal		RM	
407.6625	Ch. 5/6	FEMA 5/6	Government	P25	Federal	421 NAC	RM	
409.0625	Ch. 7/8	FEMA 7/8	Government	P25	Federal		RM	
410.4625	Ch. 9/10	FEMA 9/10	Government	P25	Federal	421 NAC	RM	

M: FRS (Family Radio Service)

FRS Radio is a system designated by the FCC in the United States as a license-free radio system for use by the public. FRS Radio restrictions are set to no more than 2 watts per channel on most channels, and limited to ½ watt of power on channels 8 thru 14. The frequencies used by the FRS Radio service are shared by GMRS Radios. Antennas on FRS Radios must be permanently fixed to the radio which greatly reduces radio range. All FRS channels are "SIMPLEX" channels restricting use to be between radios only and does not support radio repeaters. Most FRS Radios prior to 2018 are/were sold as combination radios with GMRS radios, however with the 2017 FCC Rule changes in May 2017, no future radios will be certified by the FCC for use that are combination GMRS/FRS radios which will result in a change in how the radios are packaged and marketed.

Note: The 2-watt power restriction is new as of 2017. Prior to these FCC Changes the power was restricted to ½ watt on ALL channels.

Frequencies with a "Dot" color reference are those once designated as an Itinerant Business radio channel but have been repurposed for GMRS use. As GMRS shares these frequencies with lower power license-free FRS Radio these designations are listed here as well. Businesses which have an old valid license for use of these still have priority use of these frequencies.

Frequency	Description	Name	Band	Mode	Tone	Modulation	License	Power
462.5625	FRS and GMRS simplex	FRS Ch 1	UHF	NFM	67	Simplex	Not Needed	2 Watts
462.5875	FRS and GMRS simplex	FRS Ch 2	UHF	NFM	67	Simplex	Not Needed	2 Watts
462.6125	FRS and GMRS simplex	FRS Ch 3	UHF	NFM	67	Simplex	Not Needed	2 Watts
462.6375	FRS and GMRS simplex	FRS Ch 4	UHF	NFM	67	Simplex	Not Needed	2 Watts

Frequency	Description	Name	Band	Mode	Tone	Modulation	License	Power
462.6625	FRS and GMRS simplex	FRS Ch 5	UHF	NFM	67	Simplex	Not Needed	2 Watts
462.6875	FRS and GMRS simplex	FRS Ch 6	UHF	NFM	67	Simplex	Not Needed	2 Watts
462.7125	FRS and GMRS simplex	FRS Ch 7	UHF	NFM	67	Simplex	Not Needed	2 Watts
467.5625	FRS and GMRS simplex	FRS Ch 8	UHF	NFM	67	Simplex	Not Needed	1/2 Watt
467.5875	FRS and GMRS simplex	FRS Ch 9	UHF	NFM	67	Simplex	Not Needed	1/2 Watt
467.6125	FRS and GMRS simplex	FRS Ch 10	UHF	NFM	67	Simplex	Not Needed	1/2 Watt
467.6375	FRS and GMRS simplex	FRS Ch 11	UHF	NFM	67	Simplex	Not Needed	1/2 Watt
467.6625	FRS and GMRS simplex	FRS Ch 12	UHF	NFM	67	Simplex	Not Needed	1/2 Watt
467.6875	FRS and GMRS simplex	FRS Ch 13	UHF	NFM	67	Simplex	Not Needed	1/2 Watt
467.7125	FRS and GMRS simplex	FRS Ch 14	UHF	NFM	67	Simplex	Not Needed	1/2 Watt
462.5500	FRS and GMRS simplex	FRS Ch 15	UHF	NFM	67	Simplex	Not Needed	2 Watts
462.5750	FRS and GMRS simplex (White Dot)	FRS Ch 16	UHF	NFM	67	Simplex	Not Needed	2 Watts
462.6000	FRS and GMRS simplex	FRS Ch 17	UHF	NFM	67	Simplex	Not Needed	2 Watts
462.6250	FRS and GMRS simplex (Black Dot)	FRS Ch 18	UHF	NFM	67	Simplex	Not Needed	2 Watts
462.6500	FRS and GMRS simplex	FRS Ch 19	UHF	NFM	67	Simplex	Not Needed	2 Watts
462.6750	FRS and GMRS simplex (Orange Dot)	FRS Ch 20	UHF	NFM	67	Simplex	Not Needed	2 Watts
462.7000	FRS and GMRS simplex	FRS Ch 21	UHF	NFM	67	Simplex	Not Needed	2 Watts
462.7250	FRS and GMRS simplex	FRS Ch 22	UHF	NFM	67	Simplex	Not Needed	2 Watts

N: GMRS (General Mobile Radio Service)

(http://securbar.com/part95.htm)

The GMRS Radio service is a LICENSED radio service available in the United States to the public. Licenses are available from the FCC without at test for a period of 10 years at a cost of approximately $70 and cover all family members.

Although GMRS radios share most of the same frequencies with FRS radio, the following are significant differences from FRS Radio:

1. GMRS is NOT a channelized radio service, but rather has assigned frequencies that do not follow an established channel pattern by the FCC. This has led to some confusion as different radio manufacturers assign channels, however channel designations do differ between manufacturers. Though all GMRS radios support the same frequencies, channel assignments may differ between manufacturers.
2. Power levels up to 5-watts on most channels with hand-held radios, 50-watts on mobile radios (Though power is limited to ½ watt on some frequencies shared with FRS.)
3. GMRS Radios are allowed removable antennas which allow for greater antenna options, and therefore greater range capability.
4. GMRS Radios can utilize radio repeaters which will also offer much greater radio range.
5. External antennas for GMRS radios can be placed up to 20 feet tall.
6. Frequencies with a "Dot" color reference are those once designated as an Itinerant Business radio channel but have been repurposed for GMRS use. As GMRS shares these frequencies with lower power license-free FRS Radio. Businesses which have an old valid license for use of these still have priority use of these frequencies.

GMRS Frequencies Chart (Continued)

Frequency	Description	Note	Mode	Tone	Modulation	License	Power
462.5500	Simplex, Repeater output	Shares FRS Ch 15				Yes	FRS 2 W/GMRS 50 W
462.5625	Simplex	Shares FRS Ch 1	NFM	67	Simplex	Yes	FRS 2 W/GMRS 5 W
462.5750	Simplex, Repeater output (White Dot)	Shares FRS Ch 16				Yes	FRS 2 W/GMRS 50 W
462.5875	Simplex	Shares FRS Ch 2	NFM	67	Simplex	Yes	FRS 2 W/GMRS 5 W
462.6000	Simplex, Repeater output	Shares FRS Ch 17				Yes	FRS 2 W/GMRS 50 W
462.6125	Simplex	Shares FRS Ch 3	NFM	67	Simplex	Yes	FRS 2 W/GMRS 5 W
462.6250	Simplex, Repeater output (Black Dot)	Shares FRS Ch 18				Yes	FRS 2 W/GMRS 50 W
462.6375	Simplex	Shares FRS Ch 4	NFM	67	Simplex	Yes	FRS 2 W/GMRS 5 W
462.6500	Simplex, Repeater output	Shares FRS Ch 19				Yes	FRS 2 W/GMRS 50 W
462.6625	Simplex	Shares FRS Ch 5	NFM	67	Simplex	Yes	FRS 2 W/GMRS 5 W
462.6750	Simplex, Repeater output	Shares FRS Ch 20				Yes	FRS 2 W/GMRS 50 W
462.6875	Simplex	Shares FRS Ch 6	NFM	67	Simplex	Yes	FRS 2 W/GMRS 5 W
462.7000	Simplex, Repeater output	Shares FRS Ch 21				Yes	FRS 2 W/GMRS 50 W
462.7125	Simplex	Shares FRS Ch 7	NFM	67	Simplex	Yes	FRS 2 W/GMRS 5 W
462.7250	Simplex, Repeater output (Orange Dot)	Shares FRS Ch 22				Yes	FRS 2 W/GMRS 50 W
467.5500	Repeater input	Repeater Input				Yes	GMRS Only - 50 W
467.5625	Simplex	Shares FRS Ch 8	NFM	67	Simplex	Yes	FRS .5 W/GMRS .5 W

Frequency	Description	Note	Mode	Tone	Modulation	License	Power
467.5750	Repeater input	Repeater Input				Yes	
467.5875	Simplex	Shares FRS Ch 9	NFM	67	Simplex	Yes	FRS .5 W/GMRS .5 W
467.6000	Repeater input	Repeater Input				Yes	
467.6125	Simplex	Shares FRS Ch 10	NFM	67	Simplex	Yes	FRS .5 W/GMRS .5 W
467.6250	Repeater input	Repeater Input				Yes	
467.6375	Simplex	Shares FRS Ch 11	NFM	67	Simplex	Yes	FRS .5 W/GMRS .5 W
467.6500	Repeater input	Repeater Input				Yes	
467.6625	Simplex	Shares FRS Ch 12	NFM	67	Simplex	Yes	FRS .5 W/GMRS .5 W
467.6750	Repeater Input	Repeater Input					
467.6875	Simplex	Shares FRS Ch 13	NFM	67	Simplex	Yes	FRS .5 W/GMRS .5 W
467.7000	Simplex	Repeater Input				Yes	
467.7125	Simplex	Shares FRS Ch 14	NFM	67	Simplex	Yes	FRS .5 W/GMRS .5 W
467.7250	Simplex	Repeater Input				Yes	

More About GMRS Frequencies – Channel differences between manufacturers

Unlike Citizens Band, MURS or FRS Radio, GMRS radio is not a channelized radio system. Instead frequencies are defined much the same way as they are defined in Amateur Ham radio. As many of the GMRS radios are packaged as combined radios with FRS and they share many of the FRS channels, this adds some confusion to the belief that GMRS Channels are defined officially by the FCC Rules. They are not. The chart below shows the GMRS Frequencies, their power levels and channels as used by several radio manufacturers.

Frequency	Name	GMRS Power	FRS Ch	Baofeng	Motorola	ICON	Midland	Color Dot
462.5500	GMRS/FRS Ch 15	50 w	15	Ch 7/GMRS-15 Ch 15 REPT15 Offset	15	1	15	
462.5625	GMRS/FRS Ch 1	5 w	1	Ch 0/GMRS-01	1	9	1	
462.5750	GMRS/FRS Ch 16	50 w	16	Ch 8/GMRS-16 Ch 16 REPT16 Offset	16	2	16	White
462.5875	GMRS/FRS Ch 2	5 w	2	Ch 1/GMRS-02	2	10	2	
462.6000	GMRS/FRS Ch 17	50 w	17	Ch 9/GMRS-17 Ch 17 REPT17 Offset	17	3	17	
462.6125	GMRS/FRS Ch 3	5 w	3	Ch 2/GMRS-03	3	11	3	
462.6250	GMRS/FRS Ch 18	50 w	18	Ch 10/GMRS-18 Ch 18 REPT18 Offset	18	4	18	Black
462.6375	GMRS/FRS Ch 4	5 w	4	Ch 3/GMRS-04	4	12	4	
462.6500	GMRS/FRS Ch 19	50 w	19	Ch 11/GMRS-19 Ch 19 REPT19 Offset	19	5	19	
462.6625	GMRS/FRS Ch 5	5 w	5	Ch 4/GMRS 05	5	13	5	

Frequency	Name	GMRS Power	FRS Ch	Baofeng	Motorola	ICON	Midland	Color Dot
462.6750	GMRS/FRS Ch 20	50 w	20	Ch 12/GMRS-20 Ch 20 REPT20 Offset	20	6	20	Orange
462.6875	GMRS/FRS Ch 6	5 w	6	Ch 5/GMRS-06	6	14	6	
462.7000	GMRS/FRS Ch 21	50 w	21	Ch 13/GMRS-21 Ch 21 REPT21 Offset	21	7	21	
462.7125	GMRS/FRS Ch 7	5 w	7	Ch 6/GMRS-07	7	15	7	
462.7250	GMRS/FRS Ch 22	50 w	22	Ch 14/GMRS-22 Ch 22 REPT22 Offset	22	8	22	
467.5500	GMRS Repeater Input							
467.5625	GMRS/FRS Ch 8	½ watt	8	(A)	8		8	
467.5750	GMRS Repeater Input							
467.5875	GMRS/FRS Ch 9	½ watt	9	(A)	9		9	
467.6000	GMRS Repeater Input							
467.6125	GMRS/FRS Ch 10	½ watt	10	(A)	10		10	
467.6250	GMRS Repeater Input							
467.6375	GMRS/FRS Ch 11	½ watt	11	(A)	11		11	
467.6500	GMRS Repeater Input							
467.6625	GMRS/FRS Ch 12	½ watt	12	(A)	12		12	
467.6750	GMRS Repeater Input							
467.6875	GMRS/FRS Ch 13	½ watt	13	(A)	13		13	
467.7000	GMRS Repeater Input							
467.7125	GMRS/FRS Ch 14	½ watt	14	(A)	14		14	
467.7250	GMRS Repeater Input							

Motorola and Midland follow the FRS Channel numbering for their channels 1-22. Baofeng and Icon numbering of channels differ from the FRS Standards and from those used by Motorola and Midland. ***An important point to stress is that if you are using radios with others using radios from different manufacturers, your channel numbers may differ and some coordination will need to be made.***

O: Ham Radio Simplex Radio Frequencies

The following are designated frequencies for use by HAM Radio for radio-to-radio operation (Not through a repeater) and use simplex frequencies. Simplex frequencies mean that the radio transmits and receives on the same frequency with no frequency offset such as used by Duplex transmission through repeaters.

Frequency	Description	Alpha Tag	Band	Mode	Tag	Type
29.6000	10m FM Simplex	10m FM Simplex	10 M	FM	HAM Calling	BM
50.1250	6m SSB Calling	6m SSB Call	6 M	USB	HAM Calling	BM
52.5250	6m FM Primary Calling	6m FM Call Primary	6 M	FM	HAM Calling	BM
52.5400	6m FM Secondary Calling	6m FM Call Sec	6 M	FM	HAM Calling	BM
144.2000	2m SSB Calling	2m SSB Call	2 M	USB	HAM Calling	BM
146.5200	2m FM Simplex Calling	2m FM Call	2 M	FM	HAM Calling	BM
222.1000	1.25m SSB & CW Calling	1.25m SSB/CW	1.25 M	USB	HAM Calling	BM
223.5000	1.25m FM Simplex Calling	1.25m FM	1.25 M	FM	HAM Calling	BM
432.1000	70cm SSB Calling	70cm SSB Call	70 cm	USB	HAM Calling	BM
446.0000	70cm FM Simplex Calling	70cm FM Simplex	70 cm	FM	HAM Calling	BM
902.1000	33cm SSB/CW Calling	33cm SSB CW	33 cm	USB	HAM Calling	BM
927.5000	33cm FM Simplex Calling	33cm FM Call	33 cm	NFM	HAM Calling	BM

1294.5000	23cm FM Simplex Calling	23cm FM Call	23 cm	FM	HAM Calling	BM
1296.1000	23cm SSB/CW Calling	23cm SSB/CW	23 cm	USB	HAM Calling	BM
2304.1000	13cm SSB/CW Simplex Calling	13cm SSB/CW	13 cm	USB	HAM Calling	BM
2305.2000	13cm FM Simplex Calling	13cm FM Simplex	13 cm	FM	HAM Calling	BM

P: International Space Station (ISS)

The following frequencies are HAM frequencies that are used by the International Space Station and the Astronauts.

Frequency	Description	Alpha Tag	Band	Mode	Tag	Type	License
145.8000	VHF Voice	ISS VHF	2 M	FM	Ham	BM	Yes
145.8000	Repeater	ISS Repeater	2 M	FM	Ham	RM	Yes
145.8250	VHF Packet	ISS VHF Packet	2 M	Telm	Ham	BM	Yes
437.5500	UHF Packet	ISS UHF Packet	70 cm	Telm	Ham	BM	Yes

Q: Itinerant Business Radio

Itinerant Business Radio frequencies are reserved for use by companies and individuals operating commercial activities; educational, philanthropic, or ecclesiastical institutions; clergy activities or hospitals, clinics, or medical associations

Frequency	Description	Alpha Tag	Mode	Tag	Tone	Type
121.9500	DirecTV MetLife Snoopy 1	DTV/Snoopy 1	AM	Business	CSQ	M
123.3250	DirecTV Direct 1 MetLife Snoopy 2	DTV/Snoopy 2	AM	Business	CSQ	M
123.5250	Five State Helicopters	5-State Helo	AM	Business	CSQ	M
136.5500	Operations Air To Air	Air Combat USA	AM	Business		
136.5750	Operations Air To Air	Air Combat USA	AM	Business		
136.9250	Operations Air To Air	Air Combat USA	AM	Business		
151.5125	Circus Operations	RBBB Circus A	NFM	Deprecated	263 DPL	M
151.6250	Goodyear Blimp	Goodyear Blimp	NFM	Business	151.4 PL	M
151.6250	Midwest Helicopter Airways	MdwstHeli151	NFM	Business	210.7 PL	M
151.6250	Erickson Air Crane	Erickson Air	NFM	Business	167.9 PL	M
151.6250	Circus Operations	RBBB Circus B	NFM	Deprecated	167.9 PL	M
151.9550	Circus Operations	RBBB Circus C	NFM	Deprecated	167.9 PL	M
153.0200	Circus Operations	RBBB Circus D	NFM	Deprecated	167.9 PL	M
153.3650	Siller Brothers	Siller Bros	NFM	Deprecated	CSQ	M
154.6000	Midwest Helicopter Airways	MdwstHeli154	NFM	Business	210.7 PL	M
469.5000	Ch 1 Primary	CDI 1	NFM	Business	331 DPL	M
469.5500	Ch 2 Alternate	CDI 2	NFM	Business	331 DPL	M

R: Marine Radio

Marine Radio covers a band of VHF Frequencies ranging from 156.0 and 174 MHz reserved for use by marine boaters. The band does not typically require a license for use based on the restrictions that are implied including use exclusively by boaters. The radio uses power of up to 25 watts providing a range of up to 60 nautical miles (111km).

Frequency	Description	Alpha Tag	Mode	Tag	Type
156.0500	01A Port Operations & Commercial New Orleans Lower	Marine 01A	FM	Transportation	BM
156.1750	63A Port Operations Commercial New Orleans Lower Mississippi	Marine 63A	FM	Transportation	BM
156.2500	05A Port Operations Houston New Orleans Seattle	Marine 05A	FM	Transportation	BM
156.2750	Ch 65A Port Operations	Marine65APrt	FM	Other	BM
156.3000	06 Intership Safety	Marine 06	FM	Transportation	BM
156.3250	Ch 66A Port Operations	Marine66APrt	FM	Other	BM
156.3500	07A Commercial	Marine 07A	FM	Transportation	BM
156.3750	Ch 67 Commercial intership bridgetobridge	Marine67 B-B	FM	Other	M
156.4000	08 Commercial Intership	Marine 08	FM	Transportation	BM
156.4250	Ch 68 NonCommercial	Marine 68	FM	Other	BM
156.4500	09 Boater Calling	Marine 09	FM	Transportation	BM
156.4750	Ch 69 NonCommercial	Marine 69	FM	Other	BM
156.5000	11 Commercial	Marine 10	FM	Transportation	BM
156.5250	Digital Selective Calling no voice	Marine 70	FM	Other	BM
156.5500	11 Commerical	Marine 11	FM	Transportation	BM

Frequency	Description	Alpha Tag	Mode	Tag	Type
156.5750	Ch 71 NonCommercial	Marine 71	FM	Other	BM
156.6000	12 Port Operations	Marine 12	FM	Transportation	BM
156.6250	Ch 72 Non-Commercial Intership only	Marine 72	FM	Other	M
156.6500	13 Intership Navigation Safety Bridges	Marine 13	FM	Transportation	BM
156.6750	Ch 73 Port Operations	Marine73Port	FM	Other	BM
156.7000	14 Port Operations	Marine 14	FM	Transportation	BM
156.7250	Ch 74 Port Operations	Marine74Port	FM	Other	BM
156.7500	15 Environmental Class C EPIRB	Marine 15	FM	Transportation	M
156.8000	16 Hailing Calling Distress	Marine 16	FM	Transportation	BM
156.8500	17 State local government maritime	Marine 17	FM	Transportation	BM
156.8750	Ch 77 Port Operations Intership only	Marine77Port	FM	Other	M
156.9000	18A Commercial	Marine 18A	FM	Transportation	BM
156.9250	Ch 78A Non-Commercial	Marine 78A	FM	Other	BM
156.9500	19A Commercial	Marine 19A	FM	Transportation	BM
156.9750	Ch 79A Commercial Non-Commercial in Great Lakes only	Marine 79A	FM	Other	BM
157.0000	20TX Port Operations	Marine 20 TX	FM	Transportation	M
157.0000	20A Port Operations	Marine 20A	FM	Transportation	BM
157.0250	Ch 80A Commercial Non-Commercial in Great Lakes only	Marine 80A	FM	Other	BM
157.0500	21A Coast Guard	Marine 21A	FM	Transportation	BM
157.0750	Ch 81A US Government environmental protection	Marine 81A	FM	Federal	BM
157.1000	22A Coast Guard Liaison Maritime Safety Information Broadcasts	Marine 22A	FM	Transportation	BM

Frequency	Description	Alpha Tag	Mode	Tag	Type
157.1250	Ch 82A US Government only	Marine 82A	FM	Federal	BM
157.1500	23A Coast Guard	Marine 23A	FM	Transportation	BM
157.1750	Ch 83A US Coast Guard only	Marine83A CG	FM	Federal	BM
157.2000	24TX Marine Operator	Marine 24 TX	FM	Transportation	M
157.2250	Ch 84 Shipshore telephone Ship TX	Marine84 Shp	FM	Other	M
157.2500	25TX Marine Operator	Marine 25 TX	FM	Transportation	M
157.2750	Ch 85 Shipshore telephone Ship TX	Marine85 Shp	FM	Other	M
157.3000	26TX Marine Operator	Marine 26 TX	FM	Transportation	M
157.3250	Ch 86 Shipshore telephone Ship TX	Marine86 Shp	FM	Other	M
157.3500	27TX Marine Operator	Marine 27 TX	FM	Transportation	M
157.3750	Ch 87 Shipshore telephone	Marine 87	FM	Other	BM
157.4000	28TX Marine Operator	Marine 28 TX	FM	Transportation	M
157.4125	Automatic Identification System 3 Ship	Marine-AIS 3	Telm	Data	M
157.4250	Ch 88A Commercial intership bridgetobridge	Marine 88A	FM	Other	M
161.6000	20RX Port Operations	Marine 20 RX	FM	Transportation	B
161.8000	24RX Marine Operator	Marine 24 RX	FM	Transportation	B
161.8250	Ch 84 Shipshore telephone Shore TX	Marine84 Shr	FM	Other	B
161.8500	25RX Marine Operator	Marine 25 RX	FM	Transportation	B
161.8750	Ch 85 Shipshore telephone Shore TX	Marine85 Shr	FM	Other	B
161.9000	26RX Marine Operator	Marine 26 RX	FM	Transportation	B
161.9250	Ch 86 Shipshore telephone Shore TX	Marine86 Shr	FM	Other	B

Frequency	Description	Alpha Tag	Mode	Tag	Type
161.9500	27RX Marine Operator	Marine 27 RX	FM	Transportation	B
161.9750	Automatic Identification System 1	Marine-AIS 1	Telm	Data	M
162.0000	28RX Marine Operator	Marine 28 RX	FM	Transportation	B
162.0125	Automatic Identification System 4 Shore	Marine-AIS 4	Telm	Data	B
162.0250	Automatic Identification System 2	Marine-AIS 2	Telm	Data	M

S: Military Radio

Frequency	Description	Alpha Tag	Mode	Tag	Type
40.5000	Military Search and Rescue	Mil SAR 40.5	FM	Military	M
126.2000	Military Tower	MilTwrCommon	AM	Aircraft	BM
138.4500	Air Force Search and Rescue	AF SAR 138.4	AM	Military	M
138.7500	Air Force Search and Rescue	AF SAR 138.7	AM	Military	M
242.4000	Army Helo Common Two FourTwo Four	ArmyHeli2424	AM	Military	BM
242.5000	Army Helo Common	ArmyHeli2425	AM	Military	BM
243.0000	EmergencyGuard UHF	MilAir Emerg	AM	Aircraft	BM
252.1000	Air Force Reserves AFRS Command Post Common	AFRS CP Comm	AM	Military	BM
252.5250	USAF Common Triple 25	USAFCommT25	AM	Military	BM
255.4000	Military Flight Service Station	Mil Flt Svc	AM	Aircraft	BM
257.8000	Military Tower Common at Civilian Airports	MilTwrComCiv	AM	Aircraft	BM
282.8000	Military Search and Rescue	Military SAR	AM	Military	M
299.5000	USAF Common Cheap Suit	USAF Comm	AM	Military	BM
300.6000	USAF Common Thirty O Six	USAFComm	AM	Military	BM
300.6500	USAF Common Air to Air	USAFCommAA	AM	Military	BM
303.0000	USAF Common ThirtyThirty Winchester	USAFComm3030	AM	Military	BM
303.0500	USAF Common Air to Air	USAFComm	AM	Military	BM
311.0000	Air Combat Command ACC Command Post Primary	ACC CP Pri	AM	Military	BM
319.4000	Air Mobility Command AMC Command Post	AMC CP	AM	Military	BM
321.0000	Air Combat Command ACC Command Post Secondary	ACC CP Sec	AM	Military	BM

Frequency	Description	Alpha Tag	Mode	Tag	Type
323.8000	Airborne Command Post	Airborne CP	AM	Military	M
333.0000	USAF Common Triple Three	USAFComm333	AM	Military	BM
333.3000	USAF Common Quad Three	USAFComm	AM	Military	BM
333.5500	USAF Common Full House	USAFCommFH	AM	Military	BM
335.5500	USAF Common Full House II	USAFCommFH2	AM	Military	BM
341.7500	USAF Air to Air	USAF A/A	AM	Military	BM
345.6000	USAF Common Straight	USAFComm	AM	Military	BM
349.4000	Air Mobility Command CP Common	AMC CP	AM	Military	BM
351.0000	USAF Common Haircut	USAFComm	AM	Military	BM
357.0000	USAF Common Magnum	USAFComm	AM	Military	BM
364.2000	NORAD Air Intercept Control Common	NORAD AICC	AM	Military	BM
383.5500	Take Charge and Move Out TACAMO Data	TACAMO Data	Telm	Data	M
384.5000	USAF Common Pistol	USAFComm	AM	Military	BM
384.5500	USAF Common Pistol 5	USAFComm	AM	Military	BM
396.8750	ISR Channel 1	ISR 01	FM	Military	M
397.1250	ISR Channel 2	ISR 02	FM	Military	M
397.1750	ISR Channel 3	ISR 03	FM	Military	M
397.3750	ISR Channel 4	ISR 04	FM	Military	M
397.4250	ISR Channel 5	ISR 05	FM	Military	M
397.4750	ISR Channel 6	ISR 06	FM	Military	M
397.5500	ISR Channel 7	ISR 07	FM	Military	M

Frequency	Description	Alpha Tag	Mode	Tag	Type
397.9500	ISR Channel 8	ISR 08	FM	Military	M
398.0500	ISR Channel 9	ISR 09	FM	Military	M
399.4250	ISR Channel 10	ISR 10	FM	Military	M
399.4750	ISR Channel 11	ISR 11	FM	Military	M
399.7250	ISR Channel 12	ISR 12	FM	Military	M
399.9250	ISR Channel 13	ISR 13	FM	Military	M
399.9750	ISR Channel 14	ISR 14	FM	Military	M

T: MURS Public Use Radio

MURS radio is a license free radio service established in 2000 by the FCC for public use. It is a 2-watt radio system which allows for the use of external/updated antennas on the radio but does not allow for use with repeaters.

Frequencies with a "Dot" color reference are those once designated as an Itinerant Business radio channel but have been repurposed for MURS use. Businesses which have an old valid license for use of these still have priority use of these frequencies.

Frequency	Description	Notes	Mode	Tone	Type	License	Power
151.8200	MURS	Common Use	NFM	67	VHF	None	2 Watts
151.8800	MURS	Common Use	NFM	67	VHF	None	2 Watts
151.9400	MURS	Common Use	NFM	67	VHF	None	2 Watts
154.5700	MURS (Blue Dot)	Common Use - Blue Dot	FM	67	VHF	None	2 Watts
154.6000	MURS (Green Dot)	Common Use - Red Dot	FM	67	VHF	None	2 Watts

U: National Disaster Medical

The National Disaster Medical System (NDMS) is a federally coordinated healthcare system and partnership of the United States Departments of Health and Human Services (HHS), Homeland Security (DHS), Defense (DOD), and Veterans Affairs (VA).

Frequency	Description	Alpha Tag	Band	Mode	Tag	Type
406.8625	Ch. 1/2 Digital	NDMS 1/2 Dig	Government	P25	Federal	RM
407.2625	Ch. 3/4 Digital	NDMS 3/4 Dig	Government	P25	Federal	RM
407.4625	Ch. 5/6 Digital	NDMS 5/6 Dig	Government	P25	Federal	RM
409.0750	Ch. 7/8 Digital	NDMS 7/8 Dig	Government	P25	Federal	RM
409.4625	Ch. 14 Digital	NDMS 14 Dig	Government	P25	Federal	M
412.8375	Ch. 9 Digital	NDMS 9 Dig	Government	P25	Federal	M
412.8625	Ch. 10 Digital	NDMS 10 Dig	Government	P25	Federal	M
412.8750	Ch. 11 Digital	NDMS 11 Dig	Government	P25	Federal	M
412.8875	Ch. 13 Digital	NDMS 13 Dig	Government	P25	Federal	M
412.9000	Ch. 12 Digital	NDMS 12 Dig	Government	P25	Federal	M

V: Public Safety and Law Enforcement

Frequency	Description	Alpha Tag	Mode	Tag	Type
406.0250	CospasSarsat Rescue Beacons Ch B	C-S Beacon B	Telm	Data	M
406.0280	CospasSarsat Rescue Beacons Ch C	C-S Beacon C	Telm	Data	M
406.0370	CospasSarsat Rescue Beacons Ch F	C-S Beacon F	Telm	Data	M
406.0400	CospasSarsat Rescue Beacons Ch G	C-S Beacon G	Telm	Data	M
406.0500	CospasSarsat Rescue Beacons A	C-S Beacon A	Telm	Data	M

W: UHF Federal Law Enforcement Interoperability

Frequency	Description	Alpha Tag	Band	Mode	Tag	Tone	Type
410.2375	NC2/IR15 - Calling	NC-2/IR-15	Government	NFM	Interop	167.9 PL	RM
410.4375	IR10/IR16 - Ad hoc assignment	IR-10/IR-16	Government	NFM	Interop	167.9 PL	RM
410.6375	IR11/IR17 - Ad hoc assignment	IR-11/IR-17	Government	NFM	Interop	167.9 PL	RM
410.8375	IR12/IR18 - SAR Incident Command	IR-12/IR-18	Government	NFM	Interop	167.9 PL	RM
413.1875	IR13 - Ad hoc assignment	IR13	Government	NFM	Interop	167.9 PL	BM
413.2125	IR14 - Interagency Convoy	IR14	Government	NFM	Interop	167.9 PL	BM
409.9875	LE10/LE16 - Tactical	LE-10/LE-16	Government	NFM	Interop	167.9 PL	RM
410.1875	LE11/LE17 - Tactical	LE-11/LE-17	Government	P25	Interop	68F NAC	RM
410.6125	LE12/LR18 - Tactical	LE-12/LE-18	Government	P25	Interop	68F NAC	RM
414.0375	LE B - Calling	LE B	Government	NFM	Interop	167.9 PL	BM
414.0625	LE13 - Tactical	LE-13	Government	P25	Interop	68F NAC	BM
414.3125	LE14 - Tactical	LE-14	Government	P25	Interop	68F NAC	BM
414.3375	LE15 - Tactical 15	LE-15	Government	P25	Interop	68F NAC	BM

X: UHF Medical

Frequency	Description	Alpha Tag	Band	Mode	Tag	Type
462.9500	Med-9	Med-9	Other	NFM	EMS Dispatch	RM
462.9563	Med-91	Med-91	Other	NFM	EMS Dispatch	RM
462.9625	Med-92	Med-92	Other	NFM	EMS Dispatch	RM
462.9688	Med-93	Med-93	Other	NFM	EMS Dispatch	BM
462.9750	Med-10	Med-10	Other	NFM	EMS Dispatch	RM
462.9813	Med-101	Med-101	Other	NFM	EMS Dispatch	RM
462.9875	Med-102	Med-102	Other	NFM	EMS Dispatch	RM
462.9938	Med-103	Med-103	Other	NFM	EMS Dispatch	RM
463.0000	Med-1	Med-1	Other	NFM	EMS-Tac	RM
463.0063	Med-11	Med-11	Other	NFM	EMS-Tac	RM
463.0125	Med-12	Med-12	Other	NFM	EMS-Tac	RM
463.0188	Med-13	Med-13	Other	NFM	EMS-Tac	RM
463.0250	Med-2	Med-2	Other	NFM	EMS-Tac	RM
463.0313	Med-21	Med-21	Other	NFM	EMS-Tac	RM
463.0375	Med-22	Med-22	Other	NFM	EMS-Tac	RM
463.0438	Med-23	Med-23	Other	NFM	EMS-Tac	RM
463.0500	Med-3	Med-3	Other	NFM	EMS-Tac	RM
463.0563	Med-31	Med-31	Other	NFM	EMS-Tac	RM
463.0625	Med-32	Med-32	Other	NFM	EMS-Tac	RM

Frequency	Description	Alpha Tag	Band	Mode	Tag	Type
463.0688	Med-33	Med-33	Other	NFM	EMS-Tac	RM
463.0750	Med-4	Med-4	Other	NFM	EMS-Tac	RM
463.0813	Med-41	Med-41	Other	NFM	EMS-Tac	RM
463.0875	Med-42	Med-42	Other	NFM	EMS-Tac	RM
463.0938	Med-43	Med-43	Other	NFM	EMS-Tac	RM
463.1000	Med-5	Med-5	Other	NFM	EMS-Tac	RM
463.1063	Med-51	Med-51	Other	NFM	EMS-Tac	RM
463.1125	Med-52	Med-52	Other	NFM	EMS-Tac	RM
463.1188	Med-53	Med-53	Other	NFM	EMS-Tac	RM
463.1250	Med-6	Med-6	Other	NFM	EMS-Tac	RM
463.1313	Med-61	Med-61	Other	NFM	EMS-Tac	RM
463.1375	Med-62	Med-62	Other	NFM	EMS-Tac	RM
463.1438	Med-63	Med-63	Other	NFM	EMS-Tac	RM
463.1500	Med-7	Med-7	Other	NFM	EMS-Tac	RM
463.1563	Med-71	Med-71	Other	NFM	EMS-Tac	RM
463.1625	Med-72	Med-72	Other	NFM	EMS-Tac	RM
463.1688	Med-73	Med-73	Other	NFM	EMS-Tac	RM
463.1750	Med-8	Med-8	Other	NFM	EMS-Tac	RM
463.1813	Med-81	Med-81	Other	NFM	EMS-Tac	RM
463.1875	Med-82	Med-82	Other	NFM	EMS-Tac	RM
463.1938	Med-83	Med-83	Other	NFM	EMS-Tac	RM

Y: UHF National Interoperability

Frequency	Description	Alpha Tag	Band	Mode	Tag	Tone	Type
453.2125	UCALL40 - Calling	UCALL40/40D	Government	NFM	Interop	156.7 PL	RM
453.4625	UTAC41 - Tactical	UTAC41/41D	Government	NFM	Interop	156.7 PL	RM
453.7125	UTAC42 - Tactical	UTAC42/42D	Government	NFM	Interop	156.7 PL	RM
453.8625	UTAC43 - Tactical	UTAC43/43D	Government	NFM	Interop	156.7 PL	RM

Z: VHF Federal Incident Response Interoperability

Frequency	Description	Alpha Tag	Band	Mode	Tag	Tone	Type
169.5375	Incident Calling	NC-1/IR-5	Government	NFM	Interop	167.9 PL	RM
170.0125	Incident Command	IR-1/IR-6	Government	NFM	Interop	167.9 PL	RM
170.4125	Medical Evacuation Control	IR-2/IR-7	Government	NFM	Interop	167.9 PL	RM
170.6875	Logistics Control	IR-3/IR-8	Government	NFM	Interop	167.9 PL	RM
173.0375	Interagency Convoy	IR-4/IR-9	Government	NFM	Interop	167.9 PL	RM

AA: USAR (Urban Search and Rescue)

Frequency	Description	Alpha Tag	Band	Mode	Tag	Tone	Type
408.8625	Ch. 1/2 Digital	USAR 1/2 Dig	Government	P25	Federal	421 NAC	RM
408.8625	Ch. 1/2 Analog	USAR 1/2 Ana	Government	NFM	Federal	141.3 PL	RM
409.2625	Ch. 3/4 Digital	USAR 3/4 Dig	Government	P25	Federal		RM
409.2625	Ch. 3/4 Analog	USAR 3/4 Ana	Government	NFM	Federal	141.3 PL	RM
409.6625	Ch. 5/6 Digital	USAR 5/6 Dig	Government	P25	Federal		RM
409.6625	Ch. 5/6 Analog	USAR 5/6 Ana	Government	NFM	Federal	141.3 PL	RM
409.8625	Ch. 7/8 Digital	USAR 7/8 Dig	Government	P25	Federal	421 NAC	RM
409.8625	Ch. 7/8 Analog	USAR 7/8 Ana	Government	NFM	Federal	141.3 PL	RM
410.2625	Ch. 9/10 Digital	USAR 9/10 D	Government	P25	Federal		RM
410.2625	Ch. 9/10 Analog	USAR 9/10 A	Government	NFM	Federal	141.3 PL	RM
410.6625	Ch. 11/12 Digital	USAR 11/12 D	Government	P25	Federal		RM
410.6625	Ch. 11/12 Analog	USAR 11/12 A	Government	NFM	Federal	141.3 PL	RM
412.8250	Ch. 13 Digital	USAR 13 Dig	Government	P25	Federal		M
412.8250	Ch. 13 Analog	USAR 13 Ana	Government	NFM	Federal	141.3 PL	M
412.8500	Ch. 14 Digital	USAR 14 Dig	Government	P25	Federal		M
412.8500	Ch. 14 Analog	USAR 14 Ana	Government	NFM	Federal	141.3 PL	M

AB: VHF Law Enforcement Interoperability

Frequency	Description	Alpha Tag	Band	Mode	Tag	Tone	Type
167.0875	Calling	LE-A/LE-1	Government	NFM	Interop	167.9 PL	RM
167.2500	Tactical 2/6	LE-2/LE-6	Government	P25	Interop	68F NAC	RM
167.7500	Tactical 3/7	LE-3/LE-7	Government	P25	Interop	68F NAC	RM
168.1125	Tactical 4/8	LE-4/LE-8	Government	P25	Interop	68F NAC	RM
168.4625	Tactical 5/9	LE-5/LE-9	Government	P25	Interop	68F NAC	RM

AC: VHF Low-Band Public Services

Frequency	Description	Alpha Tag	Band	Mode	Tag	Tone	F Type
39.4600	LLAW 1 - Law Enforcement	LLAW1/1D	Government	FM	Interop	156.7 PL	RM
39.4800	LFIRE 2 - Fire	LFIRE2/2D	Government	FM	Interop	156.7 PL	RM
45.8600	LLAW 3 - Law Enforcement	LLAW3/3D	Government	FM	Interop	156.7 PL	RM
45.8800	LFIRE 4 - Fire	LFIRE4/4D	Government	FM	Interop	156.7 PL	RM

AD: VHF National Interoperability

Frequency	Description	Alpha Tag	Band	Mode	Tag	Tone	F Type
151.1375	VTAC11/36 - Tactical/Repeater	VTAC11/36	Government	NFM	Interop	156.7 PL	RM
154.4525	VTAC12/37 - Tactical/Repeater	VTAC12/37	Government	NFM	Interop	156.7 PL	RM
155.7525	VCALL10 - Calling	VCALL10	Government	NFM	Interop	156.7 PL	BM
158.7375	VTAC13/38 - Tactical/Repeater	VTAC13/38	Government	NFM	Interop	156.7 PL	RM
158.7375	VTAC34 - Tactical/Repeater	VTAC34/34D	Government	NFM	Interop	156.7 PL	RM
159.4725	VTAC14/33 - Tactical/Repeater	VTAC14/33	Government	NFM	Interop	156.7 PL	RM
159.4725	VTAC35 - Tactical/Repeater	VTAC35/35D	Government	NFM	Interop	156.7 PL	RM
161.8500	VTAC17 - Tactical	VTAC17/17D	Government	NFM	Interop	156.7 PL	RM

AE: VHF Public Safety

It is illegal to broadcast on these channels without proper license or authorization. The details provided below should be used for monitoring purposes only.

Frequency	Description	Alpha Tag	Band	Mode	Tag	Tone	Type
154.2650	VFIRE22 - Fire	VFIRE22	Government	NFM	Interop		BM
154.2725	VFIRE24 - Fire	VFIRE24	Government	NFM	Interop		BM
154.2800	VFIRE21 - Fire	VFIRE21	Government	NFM	Interop		BM
154.2875	VFIRE25 - Fire	VFIRE25	Government	NFM	Interop		BM
154.2950	VFIRE23 - Fire	VFIRE23	Government	NFM	Interop		BM
154.3025	VFIRE26 - Fire	VFIRE26	Government	NFM	Interop		BM
155.1600	VSAR16 - Search/Rescue	VSAR16	Government	NFM	Interop	127.3 PL	BM
155.3400	VMED28 - EMS	VMED28	Government	NFM	Interop		BM
155.3475	VMED29 - EMS	VMED29	Government	NFM	Interop		BM
155.4750	VLAW31 - Law Enforcement	VLAW31	Government	NFM	Interop		BM
155.4825	VLAW32 - Law Enforcement	VLAW32	Government	NFM	Interop		BM

AF: Weather Radio (Broadcast Only – Not used for 2-way radio)

Not all radio weather channels will come in at all locations. Depending on your location you may receive one or more but not likely all channels. You will need to check your area on NOAA's website for coverage details which is at http://www.nws.noaa.gov/nwr/

Frequency	Description	Alpha Tag	Band	Mode	Tone	Type
162.4000	Weather Radio 1	NWS 162.400	Government	FM	CSQ	B
162.4250	Weather Radio 2	NWS 162.425	Government	FM	CSQ	B
162.4500	Weather Radio 3	NWS 162.450	Government	FM	CSQ	B
162.4750	Weather Radio 4	NWS 162.475	Government	FM	CSQ	B
162.5000	Weather Radio 5	NWS 162.500	Government	FM	CSQ	B
162.5250	Weather Radio 6	NWS 162.525	Government	FM	CSQ	B
162.5500	Weather Radio 7	NWS 162.550	Government	FM	CSQ	B

AG: VHF & UHF Television Channel Audio Frequencies

Ch	Freq	Band	Ch	Freq	Band	Ch	Freq	Band	Ch	Freq	Band
2	59.75 MHz	VHF Low-Band (Band I)	23	529.75 MHz	UHF	44	655.75 MHz	UHF	65	781.75 MHz	UHF
3	65.75 MHz	VHF Low-Band (Band I)	24	535.75 MHz	UHF	45	661.75 MHz	UHF	66	787.75 MHz	UHF
4	71.75 MHz	VHF Low-Band (Band I)	25	541.75 MHz	UHF	46	667.75 MHz	UHF	67	793.75 MHz	UHF
5	81.75 MHz	VHF Low-Band (Band I)	26	547.75 MHz	UHF	47	673.75 MHz	UHF	68	799.75 MHz	UHF
6	87.75 MHz	VHF Low-Band (Band I)	27	553.75 MHz	UHF	48	679.75 MHz	UHF	69	805.75 MHz	UHF
7	179.75 MHz	VHF Low-Band (Band II)	28	559.75 MHz	UHF	49	685.75 MHz	UHF	70	811.75 MHz	UHF
8	185.75 MHz	VHF Low-Band (Band II)	29	565.75 MHz	UHF	50	691.75 MHz	UHF	71	817.75 MHz	UHF
9	191.75 MHz	VHF Low-Band (Band II)	30	571.75 MHz	UHF	51	697.75 MHz	UHF	72	823.75 MHz	UHF
10	197.75 MHz	VHF Low-Band (Band II)	31	577.75 MHz	UHF	52	703.75 MHz	UHF	73	829.75 MHz	UHF
11	203.75 MHz	VHF Low-Band (Band II)	32	583.75 MHz	UHF	53	709.75 MHz	UHF	74	835.75 MHz	UHF
12	209.75 MHz	VHF Low-Band (Band II)	33	589.75 MHz	UHF	54	715.75 MHz	UHF	75	841.75 MHz	UHF
13	215.75 MHz	VHF Low-Band (Band II)	34	595.75 MHz	UHF	55	721.75 MHz	UHF	76	847.75 MHz	UHF
14	475.75 MHz	UHF	35	601.75 MHz	UHF	56	727.75 MHz	UHF	77	853.75 MHz	UHF
15	481.75 MHz	UHF	36	607.75 MHz	UHF	57	733.75 MHz	UHF	78	859.75 MHz	UHF
16	487.75 MHz	UHF	37	613.75 MHz	UHF	58	739.75 MHz	UHF	79	865.75 MHz	UHF
17	493.75 MHz	UHF	38	619.75 MHz	UHF	59	745.75 MHz	UHF	80	871.75 MHz	UHF
18	499.75 MHz	UHF	39	625.75 MHz	UHF	60	751.75 MHz	UHF	81	877.75 MHz	UHF
19	505.75 MHz	UHF	40	631.75 MHz	UHF	61	757.75 MHz	UHF	82	883.75 MHz	UHF
20	511.75 MHz	UHF	41	637.75 MHz	UHF	62	763.75 MHz	UHF	83	889.75 MHz	UHF
21	517.75 MHz	UHF	42	643.75 MHz	UHF	63	769.75 MHz	UHF			
22	523.75 MHz	UHF	43	649.75 MHz	UHF	64	775.75 MHz	UHF			

AH: Emergency Call Channels

These Emergency Call channels are a collection of emergency call frequencies that are used across various radio systems. Frequencies listed below in RED can be accessed from most dual band hand radios such as the Baofeng series of radios, though use may not be legal depending on the power and model of the radio. Power limits should be watched when communicating over frequencies with power restriction. For instance – Do not broadcast from a radio set to 5 watts on an FRS channel designated for a ½ watt power limit.

Understanding the difference between a "Simplex" channel and a "Not a Simplex" channel.
In HAM Radio a "Simplex" channel or radio frequency is when two radio users are communicating directly between themselves using the same frequency. Radio operator "A" keys his microphone and speaks, and then releases to listen for radio operator "B" to respond. When radio operator "B" responds, he keys his microphone and speaks over the exact same frequency. So both operators are sharing the same frequency.

In Duplex radio operation, this is most often used with repeaters. A repeater is programmed with an "Offset" which means they transmit on one frequency and receive on another. So, when I listen in on a repeater using frequency 442.550, I am listening to communications being broadcast out on that repeater at that frequency. When I key my microphone, the radio uses the offset frequency I have programmed into it (5 MHz for UHF, 600 kHz for VHF for instance) and transmits on the frequency of 447.550.

Half Duplex is when my radio can only transmit or receive at once. I am using two frequencies, but only one at a time.

With a Full Duplex radio, the radio can transmit on one frequency while receiving on another at the same time, just as you do when you hold a conversation in person with someone and that person keeps talking over you. You can hear your side of the conversation and theirs at the same time.

Emergency Call Frequencies (Continued)

Frequency	Description	Alpha Tag	Note	Band	Mode	Tone	Type	Power
121.5000	Aircraft Emergency and Distress (VHF Guard)	VHF Guard	Aircraft Emergency and Distress (VHF Guard)	Airband	AM		M	
121.7750	Emergency Locator Transmitter (ELT) Training Beacons	ELT Training		Airband	AM		M	
27.0650	CB Channel 09 - Emergency	CB Ch 09		Citizens Band	AM	CSQ	M	4 Watts (12 watts for SSB)
462.6125	FRS and GMRS simplex	FRS/GMRS	GMRS/FRS Ch 3	FRS/GMRS	NFM	67	Simplex	FRS 2w / GMRS 5w
146.5200	2-meter FM Simplex Calling	2m FM Call	FM Simplex Calling - 2M	2 M	FM		BM	
432.1000	70-cm SSB Calling	70cm SSB Call	SSB Calling - 70cm	70 cm	USB		BM	
156.8000	16 Hailing, Calling Distress	Marine 16	16 Hailing, Calling Distress		FM			
243.0000	Emergency Guard UHF	Mil Air Emergency	Emergency Guard UHF		AM			
151.9400	MURS Channel 3			MURS	NFM	67	VHF	2 Watts

AI: Shortwave Radio Frequencies

Following are a list of shortwave radio stations that can be heard from within the United States. Not all stations broadcast always, and stations do change so not all may be current.

Frequency	Station	Country	Location	Program Type
3185	WWRB USA - Overcomer Ministr	U. S. A.	North America	-
3195	WWRB	U. S. A.	North America	-
3215	WWRB	U. S. A.	North America	-
3223	Radio SR	Swaziland	Swaziland	-
3265	Radio Mozambique	Maputo, Mozambique	Maputo, Mozambique	-
3285	FR3	Cayenne, French Guiana	Cayenne, French Guiana	French
3300	Radio Cultural	Guatemala City, Guatemala	Guatemala City, Guatemala	Religious
3380	Radio Iris	Esmeraldas, Ecuador	Esmeraldas, Ecuador	Spanish
3396	Radio Kaduna	Kaduna, Nigeria	Kaduna, Nigeria	-
4750	Radio Bertoua	Bertoua, Cameroon	Bertoua, Cameroon	-
4755	Imo Regional Radio	Imo, Nigeria	Imo, Nigeria	-
4777	Radio/TV Gabon	Libreville, Gabon	Libreville, Gabon	French
4795	Radio Nueva America	La Paz, Bolivia	La Paz, Bolivia	Spanish
4820	Radio Paz y Bien	Ambala, Ecuador	Ambala, Ecuador	Spanish
4832	Radio Reloj	San Jose, Costa Rica	San Jose, Costa Rica	Spanish
4840	WWCR USA - Overcomer Ministr	U. S. A.	North America	-
4855	Radio Clube do Para	Belem, Brazil	Belem, Brazil	Portugese
4890	National Broadcasting Commission	Papua, New Guinea	Papua, New Guinea	-

Frequency	Station	Country	Location	Program Type
4915	Voice Kenya	Nairobi, Kenya	Nairobi, Kenya	-
4920	Australian Broadcasting Commission	Brisbane, Australia	Brisbane, Australia	-
4945	Radio Colosal	Neiva, Colombia	Neiva, Colombia	Spanish
4965	Radio Santa Fe	Bogota, Colombia	Bogota, Colombia	Spanish
4980	Ecos del Torbes	San Cristobal, Venezuela	San Cristobal, Venezuela	Spanish
5020	Solomon Islands Broadcasting Service	Honiara, Solomon Islands	Honiara, Solomon Islands	-
5050	WWRB	U. S. A.	North America	-
5057	Radio Gjirokaster	Gjirokaster, Albania	Gjirokaster, Albania	Albanian
5130	WBCQ	U. S. A.	America	-
5830	WTWW	U. S. A.	North America	-
5850	WRMI USA - Overcomer Ministr	U. S. A.	North America	-
5890	WWCR USA - Overcomer Ministr	U. S. A.	North America	-
5920	WHRI	U. S. A.	North America	-
5950	Guyana Broadcasting Service	Georgetown, Guyana	Georgetown, Guyana	-
5954	Radio Casino	Puerto Limon, Costa Rica	Puerto Limon, Costa Rica	-
5960	Radio Canada International	Montreal, Canada	Montreal, Canada	-
5980	Radio RSA	Johannesburg, South Africa	Johannesburg, South Africa	-
5990	Cuba - China R. Int'l	China	America	-
6000	R. Havana Cuba	Cuba	America	-
6005	CFCX	Montreal, Canada	Montreal, Canada	-
6020	Albania - China R. Int'l	China	North America	-
6025	Radio Malaysia Kuala	Lumpur, Malaysia	Lumpur, Malaysia	Chinese
6030	CFVP	Canada	North America	-

Frequency	Station	Country	Location	Program Type
6045	Radio Australia	Lyndhurst, Australia	Lyndhurst, Australia	-
6055	Nihon Shortwave Broadcasting Company	Tokyo, Japan	Tokyo, Japan	Japanese
6060	R. Havana Cuba	Cuba	America	-
6060	Radio Nacional	Buenos Aires, Argentina	Buenos Aires, Argentina	Spanish
6070	CFRX	Canada	North America	-
6075	Radio Sutatenza	Bogota, Colombia	Bogota, Colombia	Spanish
6090	Wld Univ Network	Anguilla	America	-
6090	Radio Luxembourg	Ville Louvigny,Luxembourg	Ville Louvigny,Luxembourg	-
6095	Polskie Radio	Warsaw, Poland	Warsaw, Poland	-
6100	R. Havana Cuba	Cuba	America	-
6105	Radio New Zealand	Wellington, New Zealand	Wellington, New Zealand	-
6145	R. Havana Cuba	Cuba	America	-
6160	CKZU	Canada	North America	-
7140	Trans World Radio	Monte Carlo, Monaco	Monte Carlo, Monaco	-
7170	Radio Noumea	Noumea, New Caledonia	Noumea, New Caledonia	French
7300	Radio Noumea	Tirana, Albania	Tirana, Albania	-
7375	R. Romania Int'l	Romania	Western North America	-
7385	WHRI	U. S. A.	North America	-
7490	WBCQ USA - Overcomer Ministr	U. S. A.	America	-
7506	WRNO	U. S. A.	America	-
7520	WWCR USA - Overcomer Ministr	U. S. A.	North America	-
7570	WRMI USA - Overcomer Ministr	U. S. A.	North America	-
7730	WRMI USA - Overcomer Ministr	U. S. A.	America	-

Frequency	Station	Country	Location	Program Type
9315	R. Cairo	Egypt	North America	-
9330	WBCQ	U. S. A.	America	-
9350	WWCR USA - Overcomer Ministr	U. S. A.	North America	-
9370	WWRB USA - Overcomer Ministr	U. S. A.	North America	-
9395	WRMI USA - RAE	Argentina	North America	-
9420	VOIRI (VO Justice	Iran	North America	-
9435	V. of Korea	North Korea	North America	-
9475	WTWW	U. S. A.	North America	-
9475	Radio Cairo	Cairo, Egypt	Cairo, Egypt	-
9505	WHRI	U. S. A.	North America	-
9515	Voice of Greece	Athens, Greece	Athens, Greece	-
9525	Radio Korea	Seoul, South Korea	Seoul, South Korea	-
9530	Spanish Foreign Radio	Madrid, Spain	Madrid, Spain	-
9535	Swiss Radio International	Berne, Switzerland	Berne, Switzerland	-
9540	Radio Prague	Prague, Czech Republic	Prague, Czech Republic	-
9570	Albania - China R. Int'l	China	North America	-
9570	Radio Bucharest	Bucharest, Romania	Bucharest, Romania	-
9575	Italian Radio and Television Service	Rome, Italy	Rome, Italy	-
9580	Cuba - China R. Int'l	China	North America	-
9610	Radio-TV Algeria	Algiers, Algeria	Algiers, Algeria	Arabic
9620	Radio Berlin International	Berlin, Germany	Berlin, Germany	-
9645	Radio Norway	Oslo, Norway	Oslo, Norway	-
9720	Radio Iran	Tehran, Iran	Tehran, Iran	Farsi

Frequency	Station	Country	Location	Program Type
9730	R. Romania Int'l	Romania	Western North America	-
9745	HCJB	Quita, Ecuador	Quita, Ecuador	-
9770	Austrian Radio	Vienna, Austria	Vienna, Austria	-
9790	Cuba - China R. Int'l	China	North America	-
9800	Radio Kiev	Kiev, Ukraine	Kiev, Ukraine	-
9835	Radio Budapest	Budapest, Hungary	Budapest, Hungary	-
9840	WHRI USA - Overcomer Ministr	U. S. A.	North America	-
9925	Germany - Mighty KBC Radio	Netherlands	North America	-
9965	R. Cairo	Egypt	North America	-
9980	WWCR USA - Overcomer Ministr	U. S. A.	North America	-
10040	Voice of Vietnam	Hanoi, Vietnam	Hanoi, Vietnam	-
11565	WRMI USA - Overcomer Ministr	U. S. A.	Western North America	-
11655	Israel Radio	Jerusalem, Israel	Jerusalem, Israel	-
11690	Radio Kuwait	Kuwait City, Kuwait	Kuwait City, Kuwait	-
11705	Radio Sweden	Stockholm, Sweden	Stockholm, Sweden	-
11710	V. of Korea	North Korea	North America	-
11720	Radio Moscow	Moscow, Russia	Moscow, Russia	-
11725	Radio Sofia	Sofia, Bulgaria	Sofia, Bulgaria	-
11745	Voice of Free China	Taipei, Taiwan	Taipei, Taiwan	-
11775	Wld Univ Network	Anguilla	America	-
11815	Radio Japan	Tokyo, Japan	Tokyo, Japan	-
11825	WRMI USA - Overcomer Ministr	U. S. A.	America	-
11825	Radio Tahiti	Papeete, Tahiti	Papeete, Tahiti	Tahitian

Frequency	Station	Country	Location	Program Type
11835	4VEH	Cap Haitien, Haiti	Cap Haitien, Haiti	-
11845	Radio Canada	Montreal, Canada	Montreal, Canada	-
11850	Deutsche Welle	Cologne, Germany	Cologne, Germany	-
11890	Voice of Chile	Santiago, Chile	Santiago, Chile	-
11900	Radio RSA	Johannesburg, South Africa	Johannesburg, South Africa	-
11910	BBC	London, England	London, England	-
11930	Radio Havana Cuba	Havana, Cuba	Havana, Cuba	-
11935	Radio Portugal	Lisbon, Portugal	Lisbon, Portugal	-
11945	Radio Beijing	Beijing, China	Beijing, China	-
11955	Voice of Turkey	Ankara, Turkey	Ankara, Turkey	-
11980	Radio Moscow	Moscow, Russia	Moscow, Russia	-
12105	WTWW	U. S. A.	North America	-
12160	WWCR USA - Overcomer Ministr	U. S. A.	North America	-
13740	Cuba - China R. Int'l	China	North America	-
13845	WWCR USA - Wld Univ Network	U. S. A.	North America	-
14400	BBC	London, England	London, England	-
15038	Saudi Arabian Broadcasting Service	Riyadh, Saudi Arabia	Riyadh, Saudi Arabia	Arabic
15084	Voice of Iran	Tehran, Iran	Tehran, Iran	Farsi
15135	Radio Moscow	Moscow, Russia	Moscow, Russia	-
15140	R. Havana Cuba	Cuba	Western North America	-
15165	HCJB	Quito, Ecuador	Quito, Ecuador	-
15190	ORU	Brussels, Belgium	Brussels, Belgium	-
15205	All India Radio	New Delhi, India	New Delhi, India	-

Frequency	Station	Country	Location	Program Type
15260	BBC	London, England	London, England	-
15265	Finnish Radio	Helsinki, Finland	Helsinki, Finland	-
15275	Radio Sweden	Stockholm, Sweden	Stockholm, Sweden	-
15305	Swiss Radio International	Berne, Switzerland	Berne, Switzerland	-
15310	Radio Japan	Tokyo, Japan	Tokyo, Japan	-
15320	Radio Australia	Melbourne, Australia	Melbourne, Australia	-
15430	Radio Mexico	Mexico City, Mexico	Mexico City, Mexico	Spanish
15440	WRMI USA - Overcomer Ministr	U. S. A.	Western North America	-
15465	Radio Pakistan	Islamabad, Pakistan	Islamabad, Pakistan	Urdu
15555	USB - WJHR	U. S. A.	North America	-
15575	KBS World R.	South Korea	North America	-
15590	R. Thailand	Thailand	North America	-
15760	WHRI USA - Overcomer Ministr	U. S. A.	Western North America	-
17720	Radio France International	Paris, France	Paris, France	-
17825	Vatican Radio	Vatican City	Vatican City	-
17860	Austrian Radio	Vienna, Austria	Vienna, Austria	-
21495	Israel Radio	Jerusalem, Israel	Jerusalem, Israel	-
21525	Radio Australia	Melbourne, Australia	Melbourne, Australia	-
21625	Israel Radio	Jerusalem, Israel	Jerusalem, Israel	-
21645	Radio France International	Paris, France	Paris, France	-
21735	Radio-TV Morocco	Rabat, Morocco	Rabat, Morocco	Arabic
25790	Radio RSA	Johannesburg, South Africa	Johannesburg, South Africa	-

About the Author

Rodney Biddle is an Amateur Extra Ham in Florida and has held a Commercial radio operator's license since 1986. He holds an Associate's degree in Electronics Engineering and has been hands-on involved with technology since 1986 primarily in the realm of computer digital electronics. Since the late 1990's Rodney's focus has been in Computer networking and software development holding certifications in the Computer Networking field. Rodney has been training in the computer software realm developing support and training materials, and providing customer support and instruction on the technologies he supports. This book is part of several resources that Rodney is developing focused on introduction to communications for newcomers to radio for the purposes of emergency and non-standard communications.

This book is part of a series of books and instructional products Rodney is developing focused on the newcomer to radio communications, emergency communications and emergency preparedness to help guide the novice in what options are available to fill their needs. His goal is to help the newcomer to radio with and understanding what types of systems work under which situations and to help clarify some of the confusion that presents itself to newcomers.

Made in the USA
Monee, IL
04 November 2026

67404963R00080